EXPERIENCE IN APPLYING IAEA PRINCIPLES FOR DESIGN SAFETY TO NEW NUCLEAR POWER PLANTS

The following States are Members of the International Atomic Energy Agency:

AFGHANISTAN	GERMANY	PALAU
ALBANIA	GHANA	PANAMA
ALGERIA	GREECE	PAPUA NEW GUINEA
ANGOLA	GRENADA	PARAGUAY
ANTIGUA AND BARBUDA	GUATEMALA	PERU
ARGENTINA	GUYANA	PHILIPPINES
ARMENIA	HAITI	POLAND
AUSTRALIA	HOLY SEE	PORTUGAL
AUSTRIA	HONDURAS	QATAR
AZERBAIJAN	HUNGARY	REPUBLIC OF MOLDOVA
BAHAMAS	ICELAND	ROMANIA
BAHRAIN	INDIA	RUSSIAN FEDERATION
BANGLADESH	INDONESIA	RWANDA
BARBADOS	IRAN, ISLAMIC REPUBLIC OF	SAINT KITTS AND NEVIS
BELARUS	IRAQ	SAINT LUCIA
BELGIUM	IRELAND	SAINT VINCENT AND
BELIZE	ISRAEL	THE GRENADINES
BENIN	ITALY	SAMOA
BOLIVIA, PLURINATIONAL	JAMAICA	SAN MARINO
STATE OF	JAPAN	SAUDI ARABIA
BOSNIA AND HERZEGOVINA	JORDAN	SENEGAL
BOTSWANA	KAZAKHSTAN	SERBIA
BRAZIL	KENYA	SEYCHELLES
BRUNEI DARUSSALAM	KOREA, REPUBLIC OF	SIERRA LEONE
BULGARIA	KUWAIT	SINGAPORE
BURKINA FASO	KYRGYZSTAN	SLOVAKIA
BURUNDI	LAO PEOPLE'S DEMOCRATIC	SLOVENIA
CAMBODIA	REPUBLIC	SOUTH AFRICA
CAMEROON	LATVIA	SPAIN
CANADA	LEBANON	SRI LANKA
CENTRAL AFRICAN	LESOTHO	SUDAN
REPUBLIC	LIBERIA	SWEDEN
CHAD	LIBYA	SWITZERLAND
CHILE	LIECHTENSTEIN	SYRIAN ARAB REPUBLIC
CHINA	LITHUANIA	TAJIKISTAN
COLOMBIA	LUXEMBOURG	THAILAND
COMOROS	MADAGASCAR	TOGO
CONGO	MALAWI	TONGA
COSTA RICA	MALAYSIA	TRINIDAD AND TOBAGO
CÔTE D'IVOIRE	MALI	TUNISIA
CROATIA	MALTA	TÜRKİYE
CUBA	MARSHALL ISLANDS	TURKMENISTAN
CYPRUS	MAURITANIA	UGANDA
CZECH REPUBLIC	MAURITIUS	UKRAINE
DEMOCRATIC REPUBLIC	MEXICO	UNITED ARAB EMIRATES
OF THE CONGO	MONACO	UNITED KINGDOM OF
DENMARK	MONGOLIA	GREAT BRITAIN AND
DJIBOUTI	MONTENEGRO	NORTHERN IRELAND
DOMINICA	MOROCCO	UNITED REPUBLIC
DOMINICAN REPUBLIC	MOZAMBIQUE	OF TANZANIA
ECUADOR	MYANMAR	UNITED STATES OF AMERICA
EGYPT	NAMIBIA	URUGUAY
EL SALVADOR	NEPAL	UZBEKISTAN
ERITREA	NETHERLANDS	VANUATU
ESTONIA	NEW ZEALAND	VENEZUELA, BOLIVARIAN
ESWATINI	NICARAGUA	REPUBLIC OF
ETHIOPIA	NIGER	VIET NAM
FIJI	NIGERIA	YEMEN
FINLAND	NORTH MACEDONIA	ZAMBIA
FRANCE	NORWAY	ZIMBABWE
GABON	OMAN	
GEORGIA	PAKISTAN	

The Agency's Statute was approved on 23 October 1956 by the Conference on the Statute of the IAEA held at United Nations Headquarters, New York; it entered into force on 29 July 1957. The Headquarters of the Agency are situated in Vienna. Its principal objective is "to accelerate and enlarge the contribution of atomic energy to peace, health and prosperity throughout the world".

IAEA-TECDOC-2021

EXPERIENCE IN APPLYING IAEA PRINCIPLES FOR DESIGN SAFETY TO NEW NUCLEAR POWER PLANTS

INTERNATIONAL ATOMIC ENERGY AGENCY
VIENNA, 2023

COPYRIGHT NOTICE

For further information on this publication, please contact:

Safety Assessment Section
International Atomic Energy Agency
Vienna International Centre
PO Box 100
1400 Vienna, Austria
Email: Official.Mail@iaea.org

© IAEA, 2023
Printed by the IAEA in Austria
April 2023

IAEA Library Cataloguing in Publication Data

Names: International Atomic Energy Agency.
Title: Experience in applying IAEA principles for design safety to new nuclear power plants / International Atomic Energy Agency.
Description: Vienna : International Atomic Energy Agency, 2023. | Series: IAEA TECDOC series, ISSN 1011–4289 ; no. 2021 | Includes bibliographical references.
Identifiers: IAEAL 23-01588 | ISBN 978–92–0–118623–2 (paperback : alk. paper) | ISBN 978–92–0–118523–5 (pdf)
Subjects: LCSH: Nuclear power plants — Design and construction — Safety measures. | Nuclear power plants — Safety regulations. | Nuclear power plants — Management.

FOREWORD

IAEA Safety Standards Series No. SSR-2/1, Safety of Nuclear Power Plants: Design, was published in 2012, superseding IAEA Safety Standards Series No. NS-R-1, Safety of Nuclear Power Plants: Design, issued in 2000. Among the most significant changes to SSR-2/1 compared with NS-R-1 were the inclusion of design extension conditions in the plant states considered for the design and strengthened independence of levels of defence in depth. In addition, SSR-2/1 also contained requirements for the mitigation of severe accidents and the 'practical elimination' of event sequences which could lead to early or large radioactive releases.

IAEA Safety Standards Series No. SSR-2/1 (Rev. 1) was published in 2016, incorporating the lessons from the Fukushima Daiichi accident that occurred in March 2011, with amendments in the following areas: further strengthening the independence of levels of defence in depth; prevention of cliff edge effects, including in relation to external hazards; reinforcement of capabilities for the heat transfer to the ultimate heat sink; and provisions for facilitating the use of non-permanent equipment for accident management.

The present publication aims at contributing to a common understanding of practices in Member States in the consideration of requirements established in SSR-2/1 (Rev. 1) for the design of new nuclear power plants.

To achieve this goal, Member States' experience in the consideration of new IAEA requirements for the design and safety demonstration of new nuclear power plants was collected in this publication. The publication highlights common approaches and captures potential differences pertaining to the interpretation of the requirements, or in their implementation, in the design of new nuclear power plants.

The IAEA wishes to thank the experts from Member States involved in the drafting and review of this publication for their valuable contributions. The IAEA officer responsible for this publication was S. Massara of the Division of Nuclear Installation Safety.

CONTENTS

1. INTRODUCTION

1.1. BACKGROUND

IAEA Safety Standards Series No. SSR-2/1, Safety of Nuclear Power Plants: Design, was published in 2012, superseding IAEA Safety Standards Series No. NS-R-1, Safety of Nuclear Power Plants: Design (issued in 2000). Among the most significant changes of SSR-2/1 as compared to NS-R-1 were the inclusion of design extension conditions in the plant states considered for the design and the strengthened independence of the different levels of defence in depth (DiD). In addition, SSR-2/1 also contains requirements for the mitigation of severe accidents and the 'practical elimination' of event sequences which may lead to early or large radioactive releases.

SSR-2/1 was approved by the IAEA Commission on Safety Standards a few weeks after the accident occurred at the Tokyo Electric Power Company (TEPCO) Fukushima Daiichi nuclear power plant (NPP) in March 2011. The revision of SSR-2/1 [1] (jointly with the revision of other IAEA Safety Requirements publications: IAEA Safety Standards Series No. GSR Part 1 (Rev. 1), Governmental, Legal and Regulatory Framework for Safety [2], No. GSR Part 4 (Rev. 1), Safety Assessment for Facilities and Activities [3], No. SSR-2/2 (Rev. 1), Safety of Nuclear Power Plants: Commissioning and Operation [4], and No. NS-R-3 (Rev. 1), superseded by IAEA Specific Safety Requirements No. SSR-1, Site Evaluation for Nuclear Installations [5]), was initiated in October 2012, aiming at incorporating lessons learned from the accident. The eventual review of the safety requirements had "confirmed so far the adequacy of the current safety requirements. The review revealed no significant areas of weakness, and just a small set of amendments were proposed to strengthen the requirements and facilitate their implementation"[1] [1].

SSR-2/1 (Rev. 1) [1] was finally published in 2016, with amendments in the following areas:

- Further strengthening the independence of levels of DiD;
- Prevention of cliff edge effects, including in relation to external hazards;
- Reinforcement of capabilities for the heat transfer to the ultimate heat sink (UHS);
- Provisions for facilitating the use of non-permanent equipment (NPE) for accident management.

The amended requirements in SSR-2/1 (Rev. 1) [1] and in the other relevant Safety Requirements publications were further reflected in a revision of supporting safety guides for general design aspects and for the design of specific plant systems. With this revision process almost approaching completion, the current publication aims at contributing to a common understanding of practices in Member States in the consideration of safety requirements established in SSR-2/1 (Rev. 1) [1] for the design of new nuclear power plants (NPPs).

1.2. OBJECTIVE

The purpose of this publication is to collect practical experience in different Member States in the consideration of the new IAEA principles for design safety and safety demonstration of new NPPs, as established in SSR-2/1 (Rev. 1) [1] and supporting safety guides.

---

[1] Preface of IAEA Specific Safety Requirements No. SSR-2/1 (Rev. 1), Safety of Nuclear Power Plants: Design [1].

1.3. SCOPE

This publication summarizes the practical experience of various Member States in the consideration of SSR-2/1 (Rev. 1) [1] requirements for the design and safety demonstration of new NPPs, covering in particular the following topics:

- Consideration of design extension conditions (DECs) in the categories of plant states;
- Strengthened requirements in relation to the implementation of DiD and its independence;
- 'Practical elimination' of conditions that could lead to early or large radioactive releases;
- The specific need for margins to avoid cliff edge effects;
- Strengthened requirements for the design against external hazards;
- Reinforcement of capabilities for the heat transfer to the UHS;
- Supplementary features to allow the use of NPE for accident management.

1.4. STRUCTURE

The TECDOC consists of eight main sections — referred to as the main body — completed by an appendix and a set of annexes.

Section 1 describes the background, objectives, scope and structure of the publication.

Section 2 discusses the categories of plant states considered for the design of new NPPs, with a focus on DECs. For each of the two categories of DECs, practices in their identification, analysis and in the definition of additional safety features for DECs are illustrated.

Section 3 focuses on practices in the implementation of requirements for strengthening the independence of levels of DiD.

Section 4 focuses on the 'practical elimination' of conditions that could lead to early or large radioactive releases. The scope and conditions submitted to a demonstration of 'practical elimination' are presented, together with an illustration of the demonstration for various scenarios (including specific design features identified or credited for the specific purpose of the demonstration of 'practical elimination').

Section 5 focuses on practices in the consideration of requirements demonstrating the presence of adequate margins to avoid cliff edge effects, pertaining to internal events and failures of structures, systems and components (SSCs).

Section 6 presents practices in the implementation of strengthened requirements for the design against external hazards in the design of new NPPs, covering both design basis external hazards as well as levels of natural hazards exceeding those considered for design. For each of the two categories, Member States' practices for the design as well as for the safety assessment are summarized: identification of external events, definition of levels of magnitude, list of SSCs designed or protected, and methodologies, rules and assumptions.

Section 7 summarizes practices in the reinforcement of capabilities for the heat transfer to the UHS. The section presents examples of national regulations and illustrates approaches adopted by reactor designers to ensure the necessary redundancy and diversity of heat transfer pathways to the UHS.

Section 8 covers the requirements requesting the provision of features to allow the use of NPE for accident management, for evacuating the heat from the nuclear fuel and the containment, and to provide power supply when permanent sources implemented in the design have become unavailable.

The Appendix illustrates, in addition to the practices presented in Section 4, detailed examples of the demonstration of the 'practical elimination' of conditions that could lead to early or large radioactive releases.

The Annexes illustrate individual Member States' approaches and practices in the application of IAEA principles for design safety to new NPPs. No attempt was made to harmonize the format or content; as such, the annexes reflect the Member States' specific approaches and experiences.

2. CONSIDERATION OF DESIGN EXTENSION IN THE CATEGORIES OF PLANT STATES

Subsection 2.1 presents an overview of related requirements in SSR-2/1 (Rev.1) [1] while subsection 2.2 illustrates the experience in their implementation for the design and safety demonstration of new NPPs.

2.1. REQUIREMENTS FROM SSR-2/1 (REV. 1)

Requirements introducing the categories of plant states — in particular DECs that are the focus of this section — considered for the plant design are summarized hereafter:

- SSR-2/1 (Rev. 1) [1] Section 4 (principal technical requirements): Requirement 4 (fundamental safety functions), Requirement 7 (application of defence in depth);

- SSR-2/1 (Rev. 1) [1] Section 5 (general plant design): Requirement 13 (categories of plant states), Requirement 16 (postulated initiating events), Requirement 19 (design basis accidents (DBAs)), Requirement 33 (safety systems, and safety features for design extension conditions, of units of a multiple unit nuclear power plant), Requirement 20 (design extension conditions), which states that:

"a set of design extension conditions shall be derived on the basis of engineering judgement, deterministic assessments and probabilistic assessments for the purpose of further improving the safety of the nuclear power plant by enhancing the plant's capabilities to withstand, without unacceptable radiological consequences, accidents that are either more severe than design basis accidents or that involve additional failures. These design extension conditions shall be used to identify the additional accident scenarios to be addressed in the design and to plan practicable provisions for the prevention of such accidents or mitigation of their consequences";

- SSR-2/1 (Rev. 1) [1] Section 6 (design of specific plant systems): Requirement 68 (design for withstanding the loss of off-site power).

2.2. EXPERIENCE IN THE IMPLEMENTATION OF SSR-2/1 (REV. 1) REQUIREMENTS

Subsection 2.2.1 recalls the definition of plant states provided in SSR-2/1 (Rev. 1) [1], while the following subsections illustrate practices in the implementation of requirements in normal operation as well as accidental conditions.

2.2.1. Definition of plant states

Requirement 13 of SSR-2/1 (Rev. 1) [1] defines the plant states that have to be considered in the design of NPPs. As shown in Table 1, these include operational states (normal operation, anticipated operational occurrences (AOOs)), and accident conditions (DBAs and DECs).

These defined plant states are largely used for the design of new NPPs, however, with some variations[2] in the exact breakdown of plant states as an application of Requirement 13.

In some countries, DECs without significant fuel degradation are referred to as DEC-A, and DECs with core melting are referred to as DEC-B. However, alternative terminology may be used, such as in Finland (see definitions in Ref. [6]) where:

- DEC comprise DEC-A (AOOs cumulated with a common cause failure (CCF)), DEC-B (AOOs cumulated with multiple failures) and DEC-C (associated to rare external hazards);
- Severe accidents constitute a separate category.

Plant states are defined on the basis of a list of postulated initiating events (PIEs), which include all foreseeable failures of SSCs, as well as operating errors and possible failures arising from internal and external hazards.

TABLE 1. PLANT STATES CONSIDERED IN DESIGN [7]

Operational states		Accident conditions		
Normal operation	Anticipated operational occurrences	Design basis accidents	Design extension conditions	
			Without significant fuel degradation	With core melting

In addition to triggering events such as AOOs and DBAs, some external hazards — such as airplane crash, earthquake or tsunami — have the capacity to generate a loss of large areas, potentially affecting a large number of SSCs. Hence, to prevent such situations through appropriate design measures, the analyses of external hazards are generally separated from the analyses of internal events, the aim being to protect a number of relevant SSCs and to demonstrate the ability to cope with such hazards[3]. Strengthened requirements for the design against external hazards are further discussed in Section 6.

The Japanese regulation (see Annex III) updated in the aftermath of the Fukushima Daiichi accident introduces the concept of 'beyond DEC', with requirements beyond the design basis domain in order to:

- Address significant NPP damage caused by extreme natural hazards, intentional airplane crashes or other malevolent acts; corresponding measures are the use of NPE as well as the recourse to a so-called 'specialized safety facility' (SSF). These measures aim at

---

[2] In Japan, for the design of the advanced boiling water reactor (ABWR), the following categories of faults are considered by Hitachi GE:
- Expected events (with a frequency $f > 10^{-2}$/r.y, equivalent to AOOs);
- Foreseeable events (with $f > 10^{-3}$ /r.y, correspondent to a part of DBAs);
- Design basis faults (with f comprised between 10^{-5} and 10^{-3}/r.y, correspondent to the remaining part of DBAs and a part of DECs without significant fuel degradation);
- Beyond design basis faults (with f comprised between 10^{-7} and 10^{-5}/r.y, correspondent to a part of DECs without significant fuel degradation);
- Severe accidents (correspondent to DECs with fuel melt).

[3] The interaction between external hazards and internal events is indeed considered, where relevant.

preventing and mitigating accidents that result in core melting or lead to containment failure and are generally considered as part of DECs in other countries[4] [8].

- Suppress dispersion of radioactive materials[5].

Additional information on the Japanese regulation can be found in Annex III.

In accordance with Requirement 13 of SSR-2/1 (Rev. 1) [1], "plant states shall be identified and shall be grouped into a limited number of categories primarily on the basis of their frequency of occurrence at the nuclear power plant". This requirement is generally fully implemented in the design of new NPPs, although with differences in the terminology and the approach to account for design and/or project specificities, or in response to specific requirements set by national regulatory bodies, in particular with regard to:

- Data availability (e.g. expected frequencies rather than operating experience frequencies);
- Complementary criteria (aiming at addressing the foreseeable impact on the plant of various considered initiating events).

Table 2 shows indicative values of the expected frequency of occurrence of individual plant states associated with PIEs. In this scheme, which is used in several countries for water cooled reactors, DBAs and DECs without significant fuel degradation overlap in the frequency range between 10^{-6} and 10^{-4}/r.y. In this range, additional drivers are considered to discriminate DBAs from DECs without significant fuel degradation; this includes considerations of the type of initiating event; for example, a single initiating event for DBAs, while rare events or combinations of events (multiple failures) are considered for DECs without significant fuel degradation.

TABLE 2. INDICATIVE EXPECTED FREQUENCY OF OCCURRENCE FOR DIFFERENT PLANT STATES (FROM REF. [9])

Plant state	Indicative expected frequency of occurrence (events per r.y)
Normal operation	–
Anticipated operational occurrence	$> 10^{-2}$
Design basis accident	$10^{-2} - 10^{-6}$
Design extension condition without significant fuel degradation	$10^{-4} - 10^{-6}$
Design extension condition with core melting	$< 10^{-6}$

Similarly, the frequency of occurrence of DECs without significant fuel degradation may be lower than the value reported in the table. This could be the case for the spent fuel pool (SFP), when located outside of the containment and where the confinement is limited by the filtering capability. In that case, to ensure appropriate measures to prevent uncovering of irradiated fuel assemblies (to prevent large radioactive releases), such less frequent situations may be

---

[4] Such as bunker systems in Germany: these systems offer an extra layer of protection for the equipment required to function during a station blackout (SBO) or loss of heat sink event, and are especially relevant for events caused by external hazards that are generally beyond the original design basis of the NPP.

[5] According to the IAEA definition (e.g. SSR-2/1 (Rev. 1) [1] para. 2.13 item 5), measures aimed at reducing radioactive releases in case of containment failure do not belong to DECs, as these are part of the level 5 of DiD (see Table 8).

considered. As the main cooling of the SFP is usually ensured by two redundant trains (as for example, the China National Nuclear Corporation (CNNC)s Advanced Pressurized water Reactor 1000 (HPR1000[6]) and Advanced Power Reactor 1000 (APR1000), see Annexes I and IV for more details on these reactor designs), a typical example of a DEC without significant fuel degradation is hence defined as a simultaneous failure of both main trains of the SFP cooling system. IAEA-TECDOC-1791, Considerations on the Application of the IAEA Safety Requirements for the Design of Nuclear Power Plants [9] identifies as means to mitigate this accident the adoption of "procedures to recover fuel cooling and to keep the fuel always submerged in water". It is a common practice to introduce an additional diverse SFP cooling system identified as a DEC safety feature to cope with this event. For the demonstration of 'practical elimination' (see Section 4), the loss of both trains and of the diversified cooling systems is considered for the APR1000.

In other countries (e.g. the United Kingdom [10]), DBAs and DECs without significant fuel degradation are considered in an overall approach aiming at reducing the risks below a targeted frequency, while the initiator of the sequence may be a single event or the result of a combination of events. In this case, a typical value for the boundary frequency for single events is 10^{-5}/r.y, whereas it may be equal to 10^{-7}/r.y for unmitigated sequences with potential for large releases.

In some countries (for instance China), the value of 10^{-7}/r.y is recommended to be adopted as a complementary judgement for the demonstration of 'practical elimination'. The topic of 'practical elimination' is further discussed in Section 4.

2.2.2. Normal operation

The term 'normal operation' refers to operation within specified operational limits and conditions, which includes the following modes of operation [7]:

- Startup;
- Power operation;
- Shutting down;
- Shutdown;
- Maintenance;
- Testing;
- Refuelling.

2.2.3. Anticipated operational occurrences

An AOO results from a deviation of an operational process from normal operation that is expected to occur at least once during the operating lifetime of a facility but which, in view of appropriate design provisions, does not cause any significant damage to items important to safety or lead to accident conditions.

These anomalies may lead to the following events:

- Increase in reactor heat removal;

---

[6] Given that two different versions of HPR1000 are developed in China, one by the China National Nuclear Corporation and one by the China General Nuclear System, the version referred to in the remainder of this publication will be explicitly indicated to as CNNC's HPR1000.

- Decrease in reactor heat removal;
- Decrease in reactor coolant system (RCS) flow rate;
- Reactivity initiated events;
- Increase in reactor coolant inventory;
- Decrease in reactor coolant inventory: very small LOCA due to the failure of an instrument line;
- Release of radioactive material from a subsystem or component: minor leakage from a radioactive waste system or fuel failure;
- Loss of normal electrical power.

2.2.4. Design basis accidents

According to Requirement 19 of SSR-2/1 (Rev. 1) [1], design basis accidents represent

> "accidents that are to be considered in the design, derived from postulated initiating events, for the purpose of establishing the boundary conditions for the nuclear power plant to withstand, without acceptable limits for radiation protection being exceeded.
>
> Design basis accidents shall be used to define the design bases, including performance criteria, for safety systems and for other items important to safety that are necessary to control design basis accident conditions, with the objective of returning the plant to a safe state and mitigating the consequences of any accidents".

This type of accident challenges the same main reactor parameters as an AOO, but with a greater intensity and/or in a shorter time as they result from more serious accident conditions including, where relevant, the failure of passive components (such as tube or pipe break), or instrumentation & control (I&C) failure.

Considering the wide range of frequencies of DBAs, these are often subdivided into two subcategories, sometimes referred to as design basis conditions (DBC) – e.g. DBC-3 and DBC-4 in Finland and France, DBA-1 and DBA-2 in the Republic of Korea – based on their expected frequency. In this case, different acceptance criteria may be defined for the analysis of these DBAs, with more stringent acceptance criteria for DBAs with higher frequencies (e.g. DBC-4 in Finland and France, and DBA-2 in the Republic of Korea).

2.2.5. Design extension conditions without significant fuel degradation

2.2.5.1. *Identification of conditions*

DECs without significant fuel degradation[7] are postulated accident conditions, not considered as DBAs, but which are generally studied in the design process in accordance with best estimate methodology.

As stated in Requirement 20 of SSR-2/1 (Rev. 1) [1],

> "a set of design extension conditions shall be derived on the basis of engineering judgement, deterministic assessments and probabilistic assessments for the purpose

---

[7] Also called DEC-A in some countries.

of further improving the safety of the nuclear power plant by enhancing the plant's capabilities to withstand, without unacceptable radiological consequences, accidents that are either more severe than design basis accidents or that involve additional failures".

The three approaches (engineering judgement, deterministic assessments, and probabilistic assessments) are generally adopted for the identification of DECs without significant fuel degradation:

- The deterministic assessments generally consider at least a combination of an AOO with failure of redundant equipment of systems used for the management of such plant states. Similarly, a combination of the most frequent DBAs with CCF of redundant equipment of safety systems used to cope with these DBAs is considered.
- Probabilistic safety assessment (PSA) is also widely used, at least to complement the list of DECs without significant fuel degradation or to screen in/out (likely/unlikely) situations, to ensure that no risk-significant event is missing after the screening by the deterministic assessment. The PSA generally includes all relevant reactor states (such as reactor at-power and at-shutdown states) as well as PIEs related to the SFP.
- Engineering judgement is used as a complement to address some of the PSA limitations that may lie, for example, in the PSA reduced scope (especially in the design phase) or in topics usually not considered in the PSA (such as long term behaviour in some accidents, combination of events, or lack of data).

In some countries, the regulatory body may require consideration of an established list of DECs without significant fuel degradation; this list is usually the result of extensive operating experience with light water reactor technology, research and development (R&D) outcomes, and a variety of risk assessment studies.

The following list of DECs without significant fuel degradation includes conditions for large pressurized water reactors (PWRs) that are widely used in various countries based on the above-mentioned approaches:

- Anticipated transient without scram (ATWS);
- Station blackout (SBO);
- Loss of UHS, or loss of access to the UHS;
- Total loss of component cooling water system;
- Total loss of feedwater supply to steam generators (SGs);
- Small break loss of coolant accident (SB-LOCA) combined with failure of the emergency core cooling system (ECCS);
- Loss of primary coolant during shutdown states (e.g. in residual heat removal (RHR) system mode);
- Loss of RHR system during shutdown states;
- Multiple steam generator tube rupture;
- Uncontrolled boron dilution;
- Partial loss of water inventory in the SFP;
- Loss of cooling of the SFP.

In addition, technology related situations (as such, not necessarily applicable to all reactor designs) are considered, with some examples indicated below:

- Large break loss of coolant accident (LB-LOCA) with failure of the active part of low pressure ECCS (for water-water energetic reactor, WWER);
- Main steam line break (MSLB) with consequential steam generator tube rupture (SGTR) (for APR1000);
- Rupture of one outboard main steam isolation valve with failure of one inboard main steam isolation valve (for the Hitachi GE ABWR design [11], [12]);
- Loss of computer based I&C:

  i In Finland and the United Kingdom, the total loss of complex computerized I&C may fall in the region of events to be considered in fault analysis, with rules and assumptions adapted to their frequency of occurrence;

  ii For the Hitachi GE ABWR, an example is the inadvertent closure of the main steam isolation valve following spurious failure of Class 1 safety system logic and control, combined with failure to activate reactor trip and safety systems.

2.2.5.2.   Identification of safety features for design extension conditions without significant fuel degradation

Paragraph 5.28 of SSR-2/1 (Rev. 1) [1] states that

> "the design extension conditions shall be used to define the design specifications for safety features and for the design of all other items important to safety that are necessary for preventing such conditions from arising or, if they do arise, for controlling them and mitigating their consequences".

Prevention and/or mitigation of DECs is expected to be achieved primarily by features implemented in the design (safety features for DECs) and not just by accident management measures. Hence, the consideration of DECs in the design requires the provision of additional equipment or an upgrade of classification and/or requirements for not classified or lower-classified equipment to reach and maintain a safe state. The use of NPE for accident management is further discussed in Section 8.

As a large part of DECs imply the failure of (all or a part of) a safety system, the mitigation strategy of DECs without significant fuel degradation generally relies on independent and diversified means to achieve the safety function(s) affected by the initiating events and/or the resulting accident sequence. This may imply either the use of diverse equipment to perform the same function or the recourse to alternative accident management route.

Examples of safety features for DECs without significant fuel degradation for various reactor designs are given in Tables 3 to 7 below.

TABLE 3. EXAMPLES OF FEATURES FOR DEC WITHOUT SIGNIFICANT FUEL DEGRADATION (DEC-A) FOR CNNC'S HPR1000

DEC-A condition	DEC-A feature
ATWS	Emergency boron injection system + Diverse actuation system
Total loss of feedwater	Passive residual heat removal system of secondary side (PRS)
SBO	SBO diesel generator (DG) + Steam driven pump of the auxiliary feedwater system (AFS) + Turbine bypass system-A
Total loss of heat sink – Power operation / Hot shutdown / Cold shutdown (major closure closed)	AFS (manual resupply of emergency feedwater tanks in the long term)
Total loss of heat sink – Cold shutdown (major closure open)	Safety injection system (SIS) with diversified cooling chain
SB-LOCA with failure of fast cooldown by SG	Feed and bleed
Main steam line break (MSLB) induced SGTR	AFS + SIS
Multiple SGTR	AFS + SIS
Low temperature overpressure with residual heat removal (RHR) safety valve failure	Pressurizer safety valve – Low pressure mode
Homogeneous boron dilution (due to failure of the isolation of the chemical and volume control system tank)	Switch suction pump of the chemical and volume control system to the in-containment refuelling water storage tank
Uncontrolled drain in mid-loop condition with refill signal failure	Manual make-up

TABLE 4. EXAMPLES OF FEATURES FOR DEC WITHOUT SIGNIFICANT FUEL DEGRADATION (DEC-A) FOR THE EUROPEAN PRESSURIZED WATER REACTOR (EPR)

DEC-A condition	Diversified strategy	Dedicated DEC-A feature
AOO + Failure of reactor trip (ATWS)	Boron injection by extra boration system (EBS)	ATWS signal (Automatic actuation of EBS)
AOO + Failure of protection system (ATWS)	Diversified protection signals	ATWS diversified reactor trip
AOO + Failure of SG cooling	Feed and bleed	Manual opening of pressurizer discharge system
Loss of off-site power (LOOP) + failure of emergency diesel generators (EDGs)	Ultimate diesel generators	Manual start of ultimate diesel generators
SB-LOCA + CCF on medium head safety injection	Injection into RCS by low head safety injection	Manual actuation of fast secondary cool down
Total loss of heat sink – Hot shutdown	Heat removal by SG	Manual resupply of emergency feedwater tanks
Total loss of heat sink – Cold shutdown (RHR system)	RCS boiling and containment cooling	Manual actuation of heat removal by the containment heat removal system (diversified cooling chain)

TABLE 5. EXAMPLES OF FEATURES FOR DEC WITHOUT SIGNIFICANT FUEL DEGRADATION (DEC-A) FOR THE ABWR

DEC-A condition	DEC-A feature
Transient with loss of high-pressure and low pressure ECCS	Alternative lower pressure core injection system
Transient with loss of high-pressure ECCS and depressurization	Alternative safety relief valves (SRVs)
Long term loss of all alternate AC power	Alternative AC power
Long term loss of all AC power with reactor core isolation cooling system (RCIC) failure	High-pressure alternate cooling system (HPAC) Alternative AC power
Long term loss of all AC power and loss of DC power	High-pressure alternate cooling system (HPAC) Alternative AC power Alternative DC power
Long term loss of all AC power with failure to reclose SRV	Alternative lower pressure core injection system Alternative AC power
Transient followed by failure of decay heat removal due to loss of water intake function	Alternative heat exchange facility
Transient followed by failure of decay heat removal due to RHR failure	Filtered containment venting system
Transient with failure of reactor scram	Alternative rod insertion Recirculation pump trip Standby liquid control system
Small and medium loss of coolant accident (LOCA): Loss of coolant accident with ECCS failure	Alternative lower pressure core injection system
Interfacing system loss of coolant accident (ISLOCA)	Blowout panel[a]

a   Safety system located in the reactor building used in case of MSLB LOCA outside the containment (DBA) to prevent the pressure containment vessel failure due to negative pressure. In case of ISLOCA, the blowout panel opens passively hence preventing the deterioration of environmental conditions in the reactor building.

12

TABLE 6. EXAMPLES OF FEATURES FOR DEC WITHOUT SIGNIFICANT FUEL DEGRADATION (DEC-A) FOR THE APR1000

DEC-A condition	DEC-A feature
ATWS due to mechanical blocking of control rods	Emergency boration system (EBS)
ATWS due to failure of the reactor protection system	Diverse protection system
SBO	Alternate AC DG
Loss of RHR	Diverse containment spray system
Total loss of spent fuel cooling	Diverse SFP cooling system
Total loss of feedwater to the SG	Feed and bleed
LOCA with loss of medium/high head safety injection	Passive AFS + Safety injection tank
Total loss of UHS during normal operations	Diverse essential service water system + Diverse component cooling water system
Multiple SGTR	Emergency SG blowdown system
MSLB with consequential steam generator tube ruptures (MSLB + SGTR)	SIS + Passive AFS

TABLE 7. EXAMPLES OF FEATURES FOR DEC WITHOUT SIGNIFICANT FUEL DEGRADATION (DEC-A) FOR THE WWER AES-2006

DEC-A condition	DEC-A feature
ATWS	Emergency boron injection system / Diverse actuation system
SBO	Passive heat removal system (PHRS) (secondary side)
Loss of UHS, or loss of access to UHS	PHRS (secondary side)
Total loss of component cooling water system	PHRS (secondary side)
Total loss of feed water supply to SGs	PHRS (secondary side)
SB-LOCA combined with failure of the ECCS	2&3 stage hydroaccumulators + PHRS (secondary side)
Multiple SGTR	Special algorithm for primary-to-secondary leaks (injection by emergency boron injection system to pressurizer + Emergency SG cooling system/PHRS (secondary side)
Uncontrolled boron dilution	Emergency boron injection system
Loss of primary coolant during shutdown states (e.g. in RHR system mode)	2&3 stage hydroaccumulators + special means for DEC management (alternative DG, alternative primary side make-up pump)
Loss of RHR system during shutdown states (reactor pressurized)	Passive heat removal system (secondary side)
Loss of RHR system during shutdown states (open reactor)	2&3 stage hydroaccumulators + special means for DEC management (alternative DG, alternative primary side make-up pump
Partial loss of water inventory in the SFP	Alternative primary side make-up pump in spend fuel injection mode
Loss of cooling of the SFP	Alternative primary side make-up pump in spend fuel injection mode or maintaining water level in the SFP by water supply SFP filling system and reagent preparation system
LB-LOCA with failure of the active part of low pressure ECCS (for WWER)	2&3 stage hydroaccumulators + PHRS (secondary side) + special means for DEC management (alternative DG, alternative intermediate circuit + alternative primary side make-up pump)
MSLB with consequential SGTR	Isolation of SG + injection by emergency boron injection system to pressurizer + emergency SG cooling system/ PHRS (secondary side)
Total loss of UHS during normal operations	PHRS (secondary side)

Once identified and designed to achieve the required performances for DEC conditions, safety features for DECs without significant fuel degradation are classified, assigned to appropriate requirements, including qualification to accidental conditions and, as far as practicable, designed to allow for periodic testing, monitoring, inspection and maintenance.

2.2.5.3. Analysis of design extension conditions without significant fuel degradation

As the frequency of occurrence of DECs without significant fuel degradation may overlap with the lowest frequency values of DBAs (or to the second category of DBAs for countries which adopt this practice, described in Section 2.2.4), acceptance criteria are in general similar, very close or in some cases identical to those of DBAs. For instance:

14

- For DECs without significant fuel degradation that might cause off-site radioactivity releases (e.g. primary to secondary leak) radioactive releases limits are generally the same as for DBAs[8];
- The limit on the fuel cladding temperature is the same — however, this is not necessarily the case for fuel failure rates;
- Similarly, the limit for the RCS pressure is the same[9] for DECs without significant fuel degradation (e.g. ATWS) that might cause overpressure of the RCS.

Specific safety limits for DECs without significant fuel degradation may be defined where no specific limit is defined for DBAs (for instance a margin to water boiling in the SFP), or in cases where limits for DBAs would not be appropriate for DECs (for instance, no uncovering of irradiated fuel assemblies in the SFP).

Computer codes used are generally best estimate, as recommended in para. 7.54 of IAEA Safety Standards Series No. SSG-2 (Rev. 1), Deterministic Safety Analysis for Nuclear Power Plants [13].

The rules and methods for analysing DECs without significant fuel degradation are generally different from those adopted for the analysis of DBAs in order to be less conservative, for example by not applying the single failure criterion (SFC)[10] (in agreement with para. 7.49 of SSG-2 (Rev. 1) [13] or by adopting less conservative rules and assumptions, for instance:

- In China, best estimate assumptions are considered. In order to demonstrate the existence of sufficient margins towards cliff edge effects, sensitivity studies are performed for those conditions which are closer to limits or have little design margin. The conservative assumptions of initial conditions used in DBA analysis can also be used for DECs without significant fuel degradation.
- In France, in order to fulfil acceptance criteria for DECs without significant fuel degradation with a high confidence level (95%), the values for the 'dominant parameters'[11] in the safety analyses are taken at a reasonably bounding value, whereas other parameters (second order of significance parameters) may be taken as their best estimate value.

2.2.6. Design extension conditions with core melting

2.2.6.1. *Identification of conditions*

Paragraph 5.27 of SSR-2/1 (Rev. 1) [1] states that:

"[these] additional safety features for design extension conditions, or this extension of the capability of safety systems, shall be such as to ensure the capability for managing accident conditions in which there is a significant amount of radioactive

---

[8] With some exceptions: for example, for the APR1000 (see Annex IV), the radiological acceptance criterion for DBA-2 is 5 mSv, while for DEC without significant fuel degradation is 10 mSv for a single event.

[9] With some exceptions: for example, for the APR1000 the RCS pressure limit is 110% for DBA-2 and 120% for DEC without significant fuel degradation.

[10] This is motivated by the unlikelihood of the DEC conditions and the number of failures already considered.

[11] The typical dominant parameters of a safety analysis are intended to be: (i) The characteristic parameters of the initial conditions; (ii) The I&C and sensors thresholds of the activated systems and components; (iii) The functional characteristic of the activated systems and components; (iv) The criteria for human actions.

material in the containment (including radioactive material resulting from severe degradation of the reactor core)".

In this context, it is a common understanding that a reactor core melt is postulated regardless of the low likelihood associated to accident sequences leading to such events, confirmed by the level 1 PSA results.

In the SFP, especially for reactor designs where it is located outside the containment and for which no mitigation measures would be effective to mitigate a radioactive release in case of significant degradation of the stored spent fuel, the possibility of a severe accident has to be kept extremely unlikely by implementing dedicated provisions; therefore, this sequence is generally considered for 'practical elimination' (see Section 4).

In accordance with Requirement 20 of SSR-2/1 (Rev. 1) [1], DECs with core melting are considered with the purpose of defining mitigating safety features specifically designed for such conditions. For this purpose, an appropriate knowledge of the phenomena associated with the different development scenarios of a severe accident is required, given that the progression of a severe accident involves a highly complex set of physical and chemical phenomena that have been the subject of extensive research programmes for decades. Today, the available knowledge provides a sound basis for designing features aimed at mitigating the consequences of these phenomena and/or at protecting the containment should such an accident occur.

When postulating DECs with core melting, accident sequences that may lead to the onset of fuel melt are clearly identified, and their frequency of occurrence evaluated by PSA or other applicable means. These sequences have the potential to induce containment failure through one or more of the following phenomena:

- Overpressurization or overheating of the containment vessel;
- High-pressure molten material release or direct heating of containment atmosphere;
- Combustion or explosion of combustible gases;
- Molten fuel coolant interaction, which may originate a steam explosion inside or outside the reactor pressure vessel (RPV);
- Molten core concrete interaction (MCCI);
- Containment bypass including SGTR.

From the list of sequences that lead to the threatened phenomena, DECs with core melting considered for the safety analysis are selected as sequences bounding the identified phenomena that may challenge the containment physical integrity. Bounding severe accident analysis cases are selected from the success path of the containment event tree considered in level 2 PSA.

2.2.6.2.   Identification of safety features for design extension conditions with core melting

The list of phenomena provided in Section 2.2.6.1 includes situations with high-pressure in the RCS, hence involving a risk of failure of the RPV or of other critical parts of the RCS. To avoid those situations, safety features for DECs with core melting in new NPP designs generally include the implementation of a reliable discharge feature to reach a low-pressure level before entering core melting conditions. As it will be further depicted in Section 4, such means allow high-pressure core melt scenarios to be 'practically eliminated.' On this basis, only low-pressure core melt situations are further considered as part of DECs with core melting in the paragraphs below.

For DECs with core melting, a specific objective of the safety analyses is to demonstrate that the plant can be brought into a state in which the containment functions can be maintained in the long term. This also implies that the cooling and stabilization of the molten fuel and the removal of heat from the containment are achieved in the long term. In this perspective, safety features for DECs with core melting include means to:

- Cool the molten fuel: for light water reactors, this may be achieved with basically two strategies:
  i. In-vessel melt retention (IVMR): if the reactor cavity is flooded before melt relocates into the lower plenum, the vessel wall would be cooled from its external surface and the outer vessel temperature would remain close to the cavity water saturation temperature. Nucleate pool boiling of the cavity water would constitute an efficient mechanism for heat removal from the molten debris in the lower plenum, hence enabling the confinement of the molten fuel inside the RPV.
  ii. Ex-vessel reactor cooling, which aims at collecting and cooling the corium outside the RPV in what is frequently called a core catcher.
- Remove the heat from the containment: this can be achieved by the provision of passive or active heat removal systems (possibly complemented by a containment filtered venting).

Another important issue of DECs with core melting is the management of combustible gases arising from the oxidation of metals, from water radiolysis, and from MCCI. Measures to manage these gases typically include:

- Recombination, using thermal or passive autocatalytic recombiners (PARs);
- Deliberate ignition, through the adoption of igniters (e.g. glow plug, spark, catalytic);
- Inerting of the containment atmosphere through an inert gas (nitrogen);
- Layout improvements, or, where the layout cannot be rearranged, means to avoid localized build-up of combustible gases.

Other features for DECs with core melting typically include:

- Filtration of leakages (through high efficiency filter, iodine filter and sand bed filter) from the primary containment collected in the secondary containment;
- Appropriate instrumentation to monitor the corium and the containment conditions.

These provisions represent design measures combined, where relevant, with operational measures.

Safety features for DECs with core melting are designed to operate as required under severe accident conditions, hence they are safety classified, with appropriate requirements, including for inspection and testing. They are also designed to withstand or be protected against external hazards (this topic is developed in Section 6).

Safety features for DECs with core melting that are located inside the containment have to be qualified to severe accident conditions. Additional details on this topic are presented in IAEA Safety Standards Series No. SSG-53, Design of the Reactor Containment and Associated Systems for Nuclear Power Plants [14].

2.2.6.3.   Analysis of design extension conditions with core melting

SSR-2/1 (Rev. 1) [1] para. 5.31A states that

> "the design shall be such that for design extension conditions, protective actions that are limited in terms of lengths of time and areas of application shall be sufficient for the protection of the public, and sufficient time shall be available to take such measures".

In practice, technical acceptance criteria are defined to ensure the integrity of the containment for all the conditions that are not 'practically eliminated'[12]. This can be achieved if the maximum acceptable loads on to the containment, such as structural limits, are not exceeded, and if a severe accident safe state (SASS)[13] can be reached.

The rules and methods for analysing DECs with core melting are generally defined on a best estimate basis with methodologies and calculation codes suitable for severe accident conditions. Recommended ways to fulfil IAEA SSR-2/1 (Rev. 1) [1] requirements are contained in SSG-2 (Rev. 1) [13] while current approaches in some Member States are extensively treated in IAEA-TECDOC-1982 [15].

SSR-2/1 (Rev. 1) [1] does not require the application of the SFC. The SFC is generally not applied for DEC analysis, with some notable exceptions (e.g. the Finnish regulation [6], which requires the SFC to be applied for controls and instrumentation for severe accident management).

While the application of the SFC is not required, the design of safety features for DECs with core melting may however consider the provision of redundant equipment reflecting upon:

- The ultimate importance of the fulfilled safety function and the consequential risk in case of a failure;
- The need for ensuring long term reliable operation, to maintain the containment integrity.

---

[12] For the 'practical elimination' of conditions potentially leading to early or large radioactive releases, see Section°4 of this TECDOC.

[13] A SASS is defined as follows: (i) The core melt is being cooled down; (ii) The decay heat is being removed from the containment; (iii) The containment integrity is maintained; (iv) Corium subcriticality is ensured.

3. STRENGTHENING THE INDEPENDENCE OF LEVELS OF DEFENCE IN DEPTH

Subsection 3.1 introduces relevant IAEA safety standards while subsection 3.2 illustrates the experience in the implementation of requirements from SSR-2/1 (Rev. 1) [1] and recommendations from related safety guides in the design and assessment of new NPPs.

3.1. RELEVANT IAEA SAFETY STANDARDS

Guidance from SSR-2/1 (Rev. 1) [1] and related safety guides is introduced in the following subsections.

3.1.1. Requirements from SSR-2/1 (Rev. 1)

Requirements related to the DiD concept in the plant design are summarized hereafter:

- SSR-2/1 (Rev. 1) [1] Section 2 (applying the safety principles and concepts): the concept of DiD is introduced in paras 2.12–2.14;

- SSR-2/1 (Rev. 1) [1] Section 4 (principal technical requirements): Requirement 7 (application of defence in depth);

- SSR-2/1 (Rev. 1) [1] Section 5 (general plant design): Requirement 20 (design extension conditions), Requirement 21 (physical separation and independence of safety systems), Requirement 22 (safety classification), Requirement 24 (common cause failure), Requirement 40 (prevention of harmful interactions of systems important to safety), Requirement 42 (safety analysis of the plant design);

- SSR-2/1 (Rev. 1) [1] Section 6 (design of specific plant systems): Requirement 46 (reactor shutdown), Requirement 64 (separation of protection systems and control systems), Requirement 68 (design for withstanding the loss of off-site power).

3.1.2. Recommendations and guidance from IAEA Safety Guides on how to comply with the safety requirements

Recommended ways to fulfil the abovementioned requirements are illustrated in IAEA Safety Guides supporting SSR-2/1 (Rev. 1) [1], see for instance:

- IAEA Safety Standards Series No. SSG-34, Design of Electrical Power Systems for Nuclear Power Plants [16];
- IAEA Safety Standards Series No. SSG-39, Design of Instrumentation and Control Systems for Nuclear Power Plants [17], paras 4.14–4.40;
- IAEA Safety Standards Series No. SSG-52, Design of the Reactor Core for Nuclear Power Plants [18], paras 2.3–2.5;
- IAEA Safety Standards Series No. SSG-53, Design of the Reactor Containment and Associated Systems for Nuclear Power Plants [14], paras 3.63–3.65;
- IAEA Safety Standards Series No. SSG-56, Design of the Reactor Coolant System and Associated Systems for Nuclear Power Plants [19], paras 3.57–3.61.

3.2. EXPERIENCE IN THE IMPLEMENTATION OF SSR-2/1 (REV. 1) REQUIREMENTS

Subsection 3.2.1 illustrates the association of levels of DiD to plant states, while subsections 3.2.2-3.2.6 illustrate various practices aiming at strengthening the independence of levels of DiD.

3.2.1. Levels of defence in depth and plant states

The concept of DiD has been used since the beginning of the nuclear era as a foundation for reactor safety design. This concept is established in IAEA Safety Standards Series No. SF-1, Fundamental Safety Principles [20]. The development of the concept in five levels is presented in para. 2.13 of SSR-2/1 (Rev. 1) [1]. In the aftermath of the Fukushima Daiichi accident, it was largely recognized that the DiD concept remains valid.

Several interpretations on the association of DiD levels to plant states exist in different countries, offering different interpretations of the DiD concept. Two of these, related to different interpretations of levels 3 and 4, already introduced and discussed in IAEA-TECDOC-1791 [9], are recalled in Table 8.

TABLE 8. LEVELS OF DEFENCE IN DEPTH FOR THE DESIGN OF NEW NUCLEAR POWER PLANTS (FROM REF. [9])

Level of defence Approach 1	Objective	Essential design means	Essential operational means	Level of defence Approach 2
Level 1	Prevention of abnormal operation and failures	Conservative design and high quality in construction of normal operation systems, including monitoring and control systems	Operational rules and normal operating procedures	Level 1
Level 2	Control of abnormal operation and detection of failures	Limitation and protection systems and other surveillance features	Abnormal operating procedures/emergency operating procedures	Level 2
Level 3 — 3a	Control of design basis accidents	Engineered safety features (safety systems)	Emergency operating procedures	Level 3
Level 3 — 3b	Control of design extension conditions to prevent core melt	Safety features for design extension conditions without core melt	Emergency operating procedures	Level 4 — 4a
Level 4	Control of design extension conditions to mitigate the consequences of severe accidents	Safety features for design extension conditions with core melting. Technical Support Centre	Complementary emergency operating procedures / severe accident management guidelines	Level 4 — 4b
Level 5	Mitigation of radiological consequences of significant releases of radioactive materials	On-site and off-site emergency response facilities	On-site and off-site emergency plans	Level 5

The two approaches differ depending on which level of DiD DECs without significant fuel degradation are associated to:

- **Approach 1**, i.e. the association of DECs without significant fuel degradation to **level 3**, has the advantage that each level has consistent clear objectives regarding the progression of the accident and the protection of the barriers, i.e. level 3 to prevent damage to the reactor core and level 4 to mitigate severe accidents for preventing off-site radioactive releases. In addition, the physical phenomena involved at each of these DiD levels are similar.
- **Approach 2**, i.e. the grouping of DECs without significant fuel degradation and of DECs with core melting in **level 4** follows the historical construction of DECs complementing the DBAs, making a noticeable difference between the set of rules for design (e.g. SFC) and safety analysis (conservative versus best estimate) to be applied for DBAs from those for DECs.

For simplicity and consistency with IAEA-TECDOC-1791 [9], the formulation of approach 1 will be used in the remainder of this section.

3.2.2. Independence in design

Each requirement for independence has its own objective that contributes to the plant safety and has to be fulfilled by adequate means with respect to the targeted objectives. As indicated from these requirements, the independence requirement — the aim of which is to provide a better NPP resilience to events — is not associated only to the DiD levels.

There exist other requirements where independence of safety items is key. For instance:

- The classification process requires that any interference between items important to safety and other items will be prevented, and that any failure of items important to safety in a lower safety class system will not propagate to a system in a higher safety class system (para. 5.35 of SSR-2/1 (Rev. 1) [1]);
- The SFC is practically implemented by the provision of redundant items in the systems necessary to manage AOOs and safety systems for DBAs. But redundancy is effective only if there is adequate independence between redundant items (Requirement 21 of SSR-2/1 (Rev. 1) [1]).

Paragraph 4.13A (and again para. 5.29 (a)) reinforces the independence of the safety features intended for DECs, especially those for DECs with core melting, from safety systems designed for DBAs.

The concept of DiD was obviously not built with the intent to achieve full independence, where the total loss of the whole set of SSCs at one level would not impair the other levels. For passive components such as the containment this would not make sense. Hence, the requirement for independence of DiD levels in SSR-2/1 (Rev. 1) [1] includes the formulation "as far as practicable" (see Requirement 7), to recognize that full independence might not be achieved.

3.2.3. Practices to ensure adequate independence

The following Sections 3.2.3.1–3.2.3.4 summarize some Member States' practices to reinforce the independence of DiD, substantiating how Requirement 7 of SSR-2/1 (Rev. 1) [1] is implemented in the design and licensing of new NPPs ("[…] The levels of defence in depth shall be independent as far as is practicable").

3.2.3.1.Placing a stronger requirement between some specific defence in depth levels

In practice, several countries have a specific requirement of independence between some of the components from a level (or a sublevel) and other levels (or sublevels), either between level 3b and previous levels, or between level 4 and previous levels.

This corresponds to the para. 4.13A of SSR-2/1 (Rev. 1) [1]. In practice, the independence requirement applies to the systems that protect the containment (cooling, hydrogen management, supports) and not to the containment itself.

The practices mentioned below are specific to a particular country:

- In China, the independence between safety features for mitigating the consequences of core melt accidents (level 4) and safety systems to mitigate DBAs (level 3a) was implemented as follows by the China Nuclear Power Engineering for the design of CNNC's HPR1000 (see Annex I for more information on this reactor design):
    i. The containment hydrogen combination system is used to control the hydrogen concentration below the safety limit during both DBAs and DECs with core melting. However, the hydrogen recombiners in the systems are independent, with two safety-classified recombiners used for DBAs and 31 separate recombiners used for DECs with core melting.
    ii. The containment spray system (CSS) can be operated during both DBAs and DECs with core melting; however, its function for DECs with core melting can be replaced by the passive containment heat removal system (PCS), which is dedicated to remove the heat from containment under both categories of DECs.
    iii. For level 3b, additional safety features which are independent from safety systems in level 3a (e.g. the passive residual heat removal system (PRS) on the secondary side) are adopted in CNNC's HPR1000 to cope with DECs without significant fuel degradation. However, for DECs without significant fuel degradation induced by PIEs together with partial failure of safety systems, other diverse system components can also be used and credited for mitigation purposes.
- In Finland, various requirements set forth by the regulatory body, the Radiation and Nuclear Safety Authority (STUK), deal with independence for systems intended for managing severe accidents:
    i. STUK's regulation [21] section 11 para. 8 requires that "the systems needed for reaching and maintaining a controlled state and the monitoring of the progress of an accident and the plant's status in severe reactor accidents in a nuclear power plant shall be independent of the systems designed for normal operation, anticipated operational occurrences and postulated accidents. The leaktightness of the containment during a severe reactor accident shall be reliably ensured".
    ii. STUK's regulation [6] requirement 431 also deals with systems for controlling severe accidents: "The systems intended for controlling severe accidents (level 4 of the defence in depth concept) shall be functionally and physically separated from the systems intended for normal operation and anticipated operational occurrences and for controlling postulated accidents and design extension conditions (levels 1, 2, 3a and 3b). The defence in depth level 4 systems intended for controlling severe reactor accidents may, for sound reasons, also be used for preventing severe core damage in design extension conditions provided that this will not undermine the ability of the systems to perform their primary function in case the conditions evolve into a severe accident".

iii. STUK's regulation [6] requirement 5219 states that "for the monitoring of severe reactor accidents, the containment shall be equipped with independent measuring and monitoring instrumentation that provides sufficient information on the progress of potential severe reactor accidents and any circumstances that may jeopardise containment integrity".

iv. STUK's regulation [6] requirement 5415 states that "the power supply (e.g. electricity, compressed air) to the systems designed for managing severe reactor accidents shall be independent of all the other power supply units and power distribution systems of the plant".

- In France:
  i. There is a requirement of independence between the systems fulfilling the safety functions under AOOs, DBAs and DECs without significant fuel degradation from those used in normal operation. The required independence is "as necessary", which may be interpreted as a reference to the top-level requirement (see SF-1 [20], para. 3.31).

  ii. The systems identified to manage DECs without significant fuel degradation should be as diversified as necessary from the safety systems used in DBA conditions, the failure of which they counteract. In this respect, particular attention has to be paid to the design of support systems.

  iii. The systems fulfilling safety functions for severe accident conditions should be, insofar as reasonably practicable, independent from the systems used during normal operation of the reactor, and from the systems and safety systems coming into play in the AOOs, DBAs or the additional safety features for DECs without significant fuel degradation. In this respect, particular attention should be paid to the design of support systems.

- In Japan:
  i. The Nuclear Regulation Authority (NRA) requires explicit independence between SSCs for DBAs, DECs, and beyond DECs (additional details are provided in Annex III).

  ii. Specifically, the specialized safety facility (SSF) introduced in Section 2 as a response to beyond DECs[14] should be as much as practicable redundant or diverse, independent from DBAs equipment and severe accident equipment. It should be protected against airplane crash[15] either by a sufficient physical distance from the reactor building (e.g. 100 m or more) or with a sufficiently robust structure. In this regard, practical requirements for the equipment to prevent the containment vessel failure in the SSF in Japan are presented in Annex III.

- In the Republic of Korea, a strong independence rule is applied between DiD levels 3b and 4 for the design of the APR1000 (see Annex IV for more details on this design):
  i. For instance, mitigatory safety features for DECs with core melting, such as the passive ex-vessel corium retaining and cooling system (PECS), emergency reactor depressurization system (ERDS) and the hydrogen mitigation system (HMS), are dedicated to DiD level 4.

---

[14] As already indicated in Section°2, there is no mention of "beyond DEC" in the IAEA safety standards; hence, this constitutes a unique feature developed in Japan in the aftermath of the Fukushima Daiichi accident.

[15] Protection of fundamental safety functions against aircraft crash is also applicable to new designs in other countries. But there is generally no such requirement (as in Japan) to provide a 'specialized safety facility' for that purpose.

ii. However, diverse systems such as the diverse CSS, diverse component cooling water system (CCWS), diverse essential chilled water system, diverse essential service water system are also used for DiD level 3b to cope with the event of a loss of RHR as well as DiD level 4 to ensure the containment integrity.

iii. The APR1000 also provides independence among mitigatory features for DECs without significant fuel degradation versus systems and safety systems used for AOOs and DBAs. However, some of the safety systems used for DBAs are also used for DECs without significant fuel degradation. This is only allowed if the corresponding DECs without significant fuel degradation does not impair the DBA systems.

3.2.3.2.   *Limitation to independence between anticipated operational occurrences and design basis accidents*

It is a common understanding that complete independence between level 2 (AOO) and level 3a (DBA) cannot be achieved in some practical cases. This mainly results in enabling use of some of the same safety systems designed for DBAs (e.g. the reactor trip) for AOOs, as they perform the same function in both levels, have sufficient redundancy, and use similar analysis rules. Nevertheless, in case of failure of such systems, for example due to CCF, an AOO would then lead directly to a DEC without significant fuel degradation.

Additionally, provisions typically in DiD level 1 are provided in some countries to reduce the frequency of AOOs, for instance control systems acting to reduce the power level, thus reducing the expectations from DiD level 2 (i.e. safety systems coping with AOOs).

In France, as part of the preparation of the enhanced version of the EPR (see Annex II for more details on this design), EDF is studying the possibility to consider AOOs as part of the level 3a and not as part of the level 2. This approach (representing a departure from the previous design) aims at strengthening the independence between DiD levels or — more realistically — to define DiD levels with the consideration of the independence requirement from the very beginning of the design process.

- DiD level 2 would have the objective of the prevention of AOOs. An example is a reactor trip, where alarm signals alert the operator in the event of failure of a non-classified component, giving an opportunity to undertake measures that keep the plant within the normal operation limits, in addition to means of automatic limitations to reduce the power level and achieve the same objective. The levels 1 and 2 aim at reducing the occurrence of initiating events requiring the actuation of safety systems.
- DiD level 3a is defined as AOOs and DBAs.

In this (new) perspective, the requirement of independence between DiD levels 2 and 3 is no longer a requirement of independence between AOOs and DBAs, which as can be seen above is generally challenging. In this proposed scheme, there would be a requirement for independence between limitations means as a part of (new) level 2 and the systems and safety systems used for AOOs and DBAs, respectively.

On the other hand, this would introduce the need for a certain independence between level 1 and level 2, where other limitations are foreseen. In order to automatically reduce the power of the plant, a separate I&C system may be designed (sensors + I&C processing), but at least some of the normal operation means will be required, for example to reduce the core power.

3.2.3.3.   Crediting SSCs from the previous level on a case-by-case basis

For a given accident scenario, equipment foreseen for previous levels of DiD may be credited, provided that it could be demonstrated that such equipment is neither lost, nor impaired directly or indirectly, during the scenario. This is mainly used for DECs without significant fuel degradation. In other words, the independence of features for DECs without significant fuel degradation from AOOs and DBAs safety systems is to be assessed scenario per scenario.

An example is the use of medium-pressure ECCS passive hydroaccumulators in new WWER plants, both for LB-LOCA (injection of coolant to the reactor, necessary for level 3a) and for SB-LOCA with a CCF on the high-head injection pumps (support to the emergency heat removal system in decreasing pressure in reactor unit, used at level 3b). There was an option to have two different systems (one dedicated to LB-LOCA and a second one for SB-LOCA),but this would have increased the complexity of the design.

3.2.3.4.   Evaluating the practicability of further independence

As highlighted in previous sections, the attempt to increase independence is subject to limitations and potential drawbacks. Seeking for increased independency should not challenge the overall balance between safety benefits and radiation protection, and any potential drawbacks (e.g. overcomplication of the design or of the plant operation). This implies examining and balancing different solutions or options in due time, considering also support systems. In a similar way, the design of specific systems dedicated to a specific level of DiD is obviously increasing the amount of auxiliary and support components requiring additional rooms to house them in the main NPP buildings, with high-level requirements and appropriate protection against hazards.

In the design of cooling water systems and heating, ventilation and air-conditioning systems, such considerations are encountered, for instance in the Republic of Korea and in the United Kingdom. The evaluation may lead either to a single system, to a system dedicated to each DiD level, or the provision of additional diversity where appropriate.

3.2.4. Independence and common cause failure

The consideration of CCF, and more generally the consideration of potential dependent failures, is presented in IAEA-TECDOC-1791 [9]. Sections 3.2.4.1 to 3.2.4.3 aim to illustrate some of the practices undertaken to prevent them.

3.2.4.1.   Role of diversity

Recognizing that the independence of two redundant components may be limited by their potential for CCF, a first step to limit such failures is to require appropriate separation between redundant components. This is generally achieved by an appropriate physical separation or isolation (mechanical and electrical interfaces) by appropriate means. This would provide appropriate protection against some events that may affect both, such as hazards, including those originating from one component (e.g. explosion, fire).

In addition, diversity constitutes an adequate answer to intrinsic CCF. Even if their likelihood of CCF can be discussed, it cannot be totally excluded that the failure of a first component may be quickly followed by the failure of a second redundant and identical component for the same common cause or the same failure mode, despite adequate physical separation in different buildings[16].

Diversity may be required by the national regulatory body to ensure independence between specific items (see Section 3.2.2). In addition to this diversity between two items or components, the need for diversity has to be assessed on their support systems or the need of support systems may be reduced. Indeed, the diversity between two pumps reduces the likelihood for their CCF, but their operation to fulfil a safety function may be weakened by another CCF if they have identical motors or are powered by identical EDGs.

A means of diversification frequently implemented in NPP design — where possible and relevant — is a combination of passive and active means/equipment or the use of reliable passive equipment[17], for example:

- Passive recombiners versus active igniters (see Section 2.2.6.2);
- Drop of control rods under gravity effect, injection of boron via a pump or an accumulator;
- Underpressure capacity or installation of an elevated water tank versus pumping;
- Recourse to natural circulation of fluids;
- Check-valves vs motor-operated valves.

3.2.4.2. *Risk of common cause failure across defence in depth levels*

The consequence of a CCF may affect several DiD levels for different reasons, such as:

- Use of the same system or the same active components in several DiD levels. The use of safety systems in normal operation is addressed in Section 3.2.4.3 below;
- Use of similar components or similar technologies in different systems belonging to different DiD levels. Avoiding such situations may prove to be particularly challenging, as the number of available and reliable technologies and suppliers (quality-certified for nuclear applications) for a given type of component may be limited. In that case, however, additional provisions may be considered (or required) for providing diversity:
  - Similar components may have different operating conditions (e.g. pressure, temperature, environment conditions, operation lifetime);
  - Manufacture provisions may be required (e.g. same supplier, but different locations, teams, supply chain);
  - Operational rules may be followed to reduce the potential for CCF (preventive maintenance on component A and B may be scheduled at different time, with different teams).

---

[16] For example, because of a weak spare component following replacement as part of preventive maintenance on both components, or because of a component reaching end of life at the same time (i.e. redundant components have similar lifetime and similar operational condition history).

[17] Passive systems are in general not sensitive to the same CCF that affect active ones, but they can be subject to other causes of failure (for instance, the sensitivity of natural circulation of water to the presence of incondensable gases, as the driving forces are lower than for active systems).

A typical example faced by several countries is related to the limited number of suppliers for diesel generators as well as the limited power and reliability of diverse electrical supplies such as gas turbines. The SBO generators may therefore be based on the same technology as the emergency generators. In such cases, adequate diversification is ensured at the design stage through differences in the technical specifications: for instance, a different power or voltage. In some countries (for instance in the Republic of Korea), the level of protection against external hazards may also differ. Provisions may also be adopted at the operational level to limit potential for CCF in the maintenance programmes and in their periodic testing.

The risk of CCF initiated by external hazards is addressed in Section 6.

3.2.4.3. *Use of safety systems in normal operation*

To ensure full independence of the safety systems, a relevant good practice in NPP design is to provide a set of independent systems located in a dedicated building, separated from the normal operation building and with separation of redundant components (separate building or at least separated areas), on standby, ready to be activated when necessary.

However, in some cases, some safety systems (or at least some components) may also be used in normal operation, for example:

- In both the EPR and the WWER, the low-pressure ECCS pump is also used for the RHR system, but in this case the residual heat and the temperature of the RCS are significantly reduced.
- Similarly, in the APR1000, the CSS shares the pumps and the heat exchangers with the shutdown cooling system, as there is no condition which uses both systems simultaneously.

In practice, the use of safety systems for normal operation is limited in new NPP designs and justified on a case-by-case basis. In the previous examples, the main justification is that these systems are only used for a limited duration as part of the shutdown of the reactor for refuelling or maintenance, making it reasonable to share existing components rather than implementing specific systems for normal operation.

3.2.5. Independence in the plant instrumentation and control

Recommendations on the design of I&C to meet the requirements established in SSR-2/1 (Rev. 1) [1] are presented in SSG-39 [17], which provides guidance for the overall I&C architecture in support of the concept of DiD applied to the design of the plant systems and in establishing DiD for the I&C system itself as a protection against CCF.

Diverse actuation systems have been implemented in various countries as a backup to the primary protection systems at existing plants, as well as in new NPP designs, to increase the reliability of safety functions in case of CCF in the protection system, in particular in complex computerized protection systems.

The increasing dependence on the software of such computerized safety systems for NPPs has increased the safety significance of potential CCF caused from the same software, used in redundant channels or redundant portions of the computerized I&C parts of safety systems. Using the same software creates a common dependency which might be sources of CCF, in addition to the CCF on identical redundant components.

Industrial standards provide good practices to reduce the risk associated to CCF in I&C [22]. However, as a consequence of different regulatory requirements or different specific licensee needs, design criteria for diverse actuation systems and the definition of what represents an 'adequate' level of diversity may be country specific.

More details on this topic are contained in IAEA-TECDOC-1848 [23] for the whole I&C, and in Ref. [24] which is focused on the software.

3.2.6. Assessment of a sufficient level of independence

The assessment of achieving 'as far as practicable' an appropriate level of independence between different DiD levels relies primarily on engineering judgement, complemented by deterministic analyses and PSA, especially where this is required by the regulatory body.

Additional details on this topic can be found in Safety Reports Series No. 46, Assessment of Defence in Depth for Nuclear Power Plants [25].

4. 'PRACTICAL ELIMINATION' OF CONDITIONS THAT COULD LEAD TO EARLY OR LARGE RADIOACTIVE RELEASES

Subsection 4.1 introduces relevant IAEA safety standards while subsection 4.2 illustrate the field of application and subsection 4.2 illustrates practices in the demonstration of the concept of 'practical elimination' for new NPPs.

4.1. RELEVANT IAEA SAFETY STANDARDS

Guidance from SSR-2/1 (Rev. 1) [1] and SSG-2 (Rev. 1) [13] is introduced in the following subsections.

4.1.1. Requirements from SSR-2/1 (Rev. 1)

Paragraph 5.27 of SSR-2/1 (Rev. 1) [1] states that "the plant shall be designed […] with the result that the possibility of plants states arising that could lead to an early radioactive release or a large radioactive release is 'practically eliminated'". Other requirements pertaining to 'practical elimination' are summarized hereafter:

- SSR-2/1 (Rev. 1) [1] Section 2 (applying the safety principles and concepts): requirements related to safety in design (para. 2.11) and the concept of defence in depth (para. 2.13 item (4)), where definitions are provided for 'early radioactive release' and 'large radioactive release'[18];
- SSR-2/1 (Rev. 1) [1] Section 4 (principal technical requirements): Requirement 5 (radiation protection in design) para. 4.3;
- SSR-2/1 (Rev. 1) [1] Section 5 (general plant design): Requirement 20 (design extension conditions) paras 5.27 and 5.31, where a definition is provided for the "practical elimination of the possibility of certain conditions arising […]"[19] [1]";
- SSR-2/1 (Rev. 1) [1] Section 6 (design of specific plant systems): Requirement 80 (fuel handling and storage system) para. 6.68.

4.1.2. Recommendations and guidance from SSG-2 (Rev. 1) on how to comply with the safety requirements

Paras 7.69–7.72 of SSG-2 (Rev. 1) [13] provide recommendations pertaining to the demonstration of 'practical elimination' of the possibility of conditions arising that could lead to an early radioactive release or a large radioactive release.

4.2. FIELD OF APPLICATION

4.2.1. Scope of 'practical elimination'

As described in previous sections, NPPs are designed in accordance with the DiD principle: a series of measures defined to prevent AOOs and accidents, and to mitigate their consequences should these accidents occur despite the measures taken to prevent them. The design mainly

---

[18] "An 'early radioactive release' in this context is a radioactive release for which off-site protective actions would be necessary but would be unlikely to be fully effective in due time. A 'large radioactive release' is a radioactive release for which off-site protective actions that are limited in terms of lengths of time and areas of application would be insufficient for the protection of people and of the environment" ([1] footnote 3).

[19] "The possibility of certain conditions arising may be considered to have been 'practically eliminated' if it would be physically impossible for the conditions to arise or if these conditions could be considered with a high level of confidence to be extremely unlikely to arise" ([1] footnote 4).

focuses on the first four levels of DiD, with the objective of ensuring the robustness of each level and an adequate independence among different levels (this topic is addressed in Section 3).

However, the implementation of DiD in the design of a NPP may encounter some limitations. Indeed, given that in NPPs large quantities of radioactive substances coexist with sufficient energy able to disperse them, there is a risk of a severe accident occurring, combined with a threat to the containment function. For such situations, the mitigation of consequences might not be possible. For instance, as seen in Section 2, it needs to be recognized that, for particular sequences of events (for example sequences with high-pressure in the RCS, involving a risk of failure of the RPV or of large portions of the RCS, see Section 2.2.6.1), the complete set of prevention and mitigation features — corresponding to all levels of DiD — might not be implemented.

Such situations are likely to lead to an early radioactive release or a large radioactive release due to the simultaneous or successive loss of integrity of all the confinement barriers or because of the bypass of these barriers (containment bypass situations) either have to:

- Lead to define provisions aiming at significantly limiting their consequences, hence allowing acceptance criteria to be met, or;
- Be 'practically eliminated' where it appears to be impossible to define such provisions or to demonstrate their adequacy with the knowledge and techniques available at the time of the design phase.

Furthermore, the concept of 'practical elimination' should not entail the lack of features allowing to prevent and mitigate accidents with core melting (as introduced in Section 2.2.6, a scenario leading to reactor core melt has to be postulated as part of DEC) or the absence of fully effective emergency arrangements both on-site and off-site, but constitutes a part of DiD. Within this framework, the concept of 'practical elimination' is understood as a means to strengthen the implementation of the concept of DiD.

SSG-2 (Rev. 1) [13] suggests grouping the cases to be addressed for 'practical elimination' within the following five categories:

a) Events that could lead to prompt reactor core damage and consequent early containment failure:
- Failure of a large component in the RCS;
- Uncontrolled reactivity accidents.
b) Severe accident phenomena which could lead to early containment failure:
- Direct containment heating (DCH);
- Large steam explosion;
- Hydrogen detonation.
c) Severe accident phenomena which could lead to late containment failure (LCF):
- Molten core concrete interaction (MCCI);
- Loss of containment heat removal.
d) Severe accident with containment bypass;
e) Significant fuel degradation in a storage pool.

The cases to be addressed for 'practical elimination' include situations where the implementation of DiD is unbalanced with one or more levels missing. For instance:

- When the failure of a large component is considered for 'practical elimination', only the level 1 measures are effective.
- When a severe accident with containment bypass is to be 'practically eliminated', only measures from levels 1 to 3 and part of level 4 are effective.
- When a significant fuel degradation in the SFP is to be 'practically eliminated', only measures from levels 1 to 3 are effective.

The position of 'practical elimination' with respect to DiD is qualitatively illustrated in Fig. 1, as a contribution to the overall objective to avoid large or early radioactive releases.

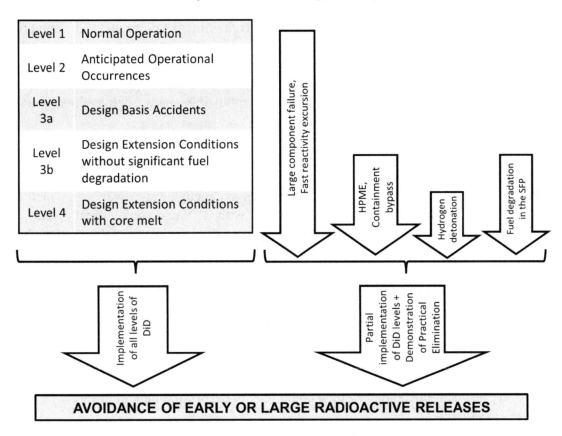

FIG. 1. Qualitative illustration of the position of 'practical elimination' and DiD for a given situation (HPME: High-Pressure Melt Ejection).

4.2.2. Conditions to be submitted to a demonstration of 'practical elimination'

4.2.2.1.   Process for the identification of conditions

This identification is made through engineering judgement, deterministic analyses, and PSA. It encompasses all reactor states (including shutdown modes) and the SFP.

The analysis of these conditions involves consideration of uncertainties due to the limited knowledge of certain physical phenomena and relying as much as necessary on dedicated studies or R&D activities.

4.2.2.2.   Examples of lists of conditions to be 'practically eliminated' for various technologies

For some of the reactor technologies addressed in this TECDOC, the conditions to be 'practically eliminated' and the design and operational provisions implemented to 'practically eliminate' the possibility of those conditions arising (see para. 7.70 of SSG-2 (Rev. 1) [13]) are presented below. In general, the list may be presented in groups like the categories recalled in Section 4.2.1. However, to better adhere to the lists of these designs, these categories are adapted as follows:

- DCH is the consequence of high-pressure melt ejection (HPME). As HPME may have other implications, for instance induced SGTR in PWRs, this is often summarized as 'HPME and DCH'.
- Steam explosion may occur in-vessel or ex-vessel and may be called fuel–coolant interaction (FCI). A limited steam explosion may be acceptable for the containment, thus only large steam explosions likely to lead to containment failure are to be 'practically eliminated'.
- It is possible to expand the term 'hydrogen detonation' to all explosion processes endangering the containment integrity (according to SSG-53 [14], hazards related to hydrogen might be broadened to include all combustible gases, such as those generated by MCCI).
- Loss of containment heat removal may be reformulated as LCF consequential to overpressure or overheating, as for instance, overpressure may be due to non-condensable gases.
- Severe accident with containment bypass may include at-power core melt with containment bypass and at-shutdown core melt with RPV and containment open.
- Significant fuel degradation in the SFP may occur in the long term following loss of cooling, or in the short term in case of failure of the SFP structure.

Thus, the five categories introduced in subsection 4.2.1 may be defined as follows:

a) Events that could lead to prompt reactor core damage and consequent early containment failure:
   - Failure of a large component in the RCS;
   - Uncontrolled reactivity accidents.
b) Severe accident phenomena which could lead to early containment failure:
   - HPME and DCH;
   - In-vessel large steam explosion;
   - Ex-vessel large steam explosion;
   - Combustion of gases endangering the containment integrity.
c) Severe accident phenomena which could lead to LCF:
   - MCCI;
   - LCF due to overpressure or overheating.
d) Severe accident with containment bypass;
   - At-power core melt with containment bypass;
   - At-shutdown core melt with RPV and containment open.
e) Significant fuel degradation in the SFP:
   - Dewatering following boiling or leakages;
   - Catastrophic failure of storage pool.

According to these categories, the list of conditions to be 'practically eliminated' for several designs are further summarized in Table 9.

TABLE 9. LIST OF CONDITIONS OBJECT OF A DEMONSTRATION OF 'PRACTICAL ELIMINATION' FOR VARIOUS REACTOR DESIGNS

Subcategory	EPR	WWER	CNNC's HPR1000	APR1000	ABWR
Failure of a large component in the RCS[a]	Yes [b]	RPV, SG, SG header	Yes	RPV, SG, pressurizer	RPV
Uncontrolled reactivity accidents	Yes	Yes	Yes	Yes	Yes
High-pressure melt ejection (HPME) and direct containment heating (DCH)	Yes	Yes	Yes	Yes	Yes
In-vessel large steam explosion	Yes	Yes	Yes	Yes	Yes
Ex-vessel large steam explosion	Yes	Yes	Yes	Yes	Yes
Combustion of gases endangering the containment integrity	Yes	Yes	Yes	Yes	Yes
Molten core concrete interaction (MCCI)	[c]	[c]	Yes	Yes	Yes
Late containment failure (LCF) due to overpressure or overheating	[c]	[c]	Yes	Yes	Yes
Severe accident with containment bypass at power	Yes	Yes	Yes	Yes	Yes
Core melt at shutdown state with RPV and containment open	Yes	Yes	Yes	Yes	Yes
Significant fuel degradation in the SFP by boil-off	Yes	Yes	Yes	Yes	Yes
Catastrophic failure of the SFP	Yes [c]	Yes	Yes	Yes	Yes

a   The 'practical elimination' of failure of a large component may include main piping of RCS and part of the main steam system.

b   The failure of a large component is not systematically considered as part of the concept of 'practical elimination' but may be part of another concept named event preclusion[20]. However, similarities exist between the application of this concept and the concept of 'practical elimination'. The difference is mainly a formal one (some publications from WENRA and OECD-NEA about 'practical elimination', for instance [26], [27], [28], include this type of sequence).

c   These sequences are considered as part of DEC with core melting (see FIG. 1, left side). Indeed, these conditions should be analysed during the identification of situations to be 'practically eliminated', even though their consequences can generally be mitigated with implementation of reasonable technical means [14].

4.3.   DEMONSTRATION OF 'PRACTICAL ELIMINATION'

4.3.1. Approaches

The demonstration of 'practical elimination' is based on a case-by-case analysis and relies on deterministic and probabilistic considerations, as well as "engineering aspects such as design, fabrication, testing and inspection of SSCs and evaluation of operating experience" (para 7.69 of SSG-2 (Rev. 1) [13]).

SSR-2/1 (Rev. 1) [1] states that "the possibility of certain conditions arising may be considered to have been 'practically eliminated' if it would be physically impossible for the conditions to arise or if these conditions could be considered with a high level of confidence to be extremely unlikely to arise". The justification of 'practical elimination' through physical impossibility of

---

[20] Or alternative wording, for instance 'incredible failure'.

the situation does not need probabilistic considerations, but para. 7.72 of SSG-2 (Rev. 1) [13] recognizes that "in practice this approach is limited to very specific cases".

The following Sections 4.3.1.1–4.3.1.3 capture common approaches by organizations having contributed to this TECDOC.

4.3.1.1. *Physically impossible conditions*

The demonstration of the physical impossibility of a situation can be based on various considerations, for example:

- Intrinsic characteristics (e.g. reactivity coefficients and other inherent safety features) allowing for exclusion of the occurrence of phenomena involved in an accident scenario;
- Design choices limiting the quantities of substances likely to initiate energetic events or phenomena (e.g. capacity limitation of unborated water tanks for the circuits connected to the reactor primary circuit to prevent heterogeneous dilutions, which would lead to a reactivity-initiated accident).

It should be noted that the demonstration of physical impossibility cannot be based on measures requiring active components, as they always present the potential for failure.

4.3.1.2. *Conditions that could be considered with a high level of confidence to be extremely unlikely to arise*

This type of justification is assessed on a case-by-case basis and is based on engineering judgement and deterministic analysis, generally complemented by probabilistic assessments, relying on one or more of the following principles:

- Design provisions relying on equipment defined within the scope of the 'practical elimination' of an accident scenario take into account requirements regarding design (such as diversification, physical separation, backup power, qualification, reliability), manufacturing (quality control) and operation (e.g. operation monitoring, periodic tests, in-service inspection). This also applies to the instrumentation used to check the functions fulfilled by these provisions. The requirements are proportionate to the expected robustness of the provision;
- The deterministic justification for a situation being 'practically eliminated' considers both the existence of sufficient lines of defence (consisting of both design measures and organizational provisions) and their robustness and independence;
- High reliability of passive static systems (such as construction provisions preventing serious damage to the structural integrity of an SFP as a result of a heavy load drop, where such damage would result in the uncovering of spent fuel assemblies or of a fuel assembly being handled);
- When the justification for 'practical elimination' is partly based on human actions (e.g. the manual opening of depressurization valves of the reactor cooling system, venting, containment isolation), provisions are provided to ensure that such actions have a low probability of failure (for instance sufficient accessibility time, clear information and procedures);
- Provisions defined within the scope of the 'practical elimination' of an accident scenario are appropriately designed or protected from internal and external hazards; in particular, the occurrence of rare and severe external natural hazards should not impair a 'practical

elimination' justification[21]. Support systems are similarly considered and, where possible, preference is generally given to the implementation of measures that are tolerant to the loss of support functions;

- The degree of substantiation provided for a demonstration of 'practical elimination' may take account of the assessed frequency of the situation to be 'practically eliminated' and of the degree of confidence in the assessed frequency (uncertainties associated with methods and data have to be evaluated to underwrite the degree of confidence claimed). Appropriate sensitivity studies may be included to confirm the existence of sufficient margin to cliff edge effects.

Such a justification is generally not solely based on compliance with a general cut-off probabilistic value, in accordance with para. 7.71 of SSG-2 (Rev. 1) [13].

4.3.1.3.   Example of probabilistic targets

When defining probabilistic criteria, only a value, or a set of values, for the overall objective of avoidance of large releases might be set in some cases, while in other cases a limiting value for individual (or a group of) sequences might be established.

IAEA-TECDOC-1791 [9] considers that, for a new design adopting the best technological solutions for a strong implementation of the concept of DiD, it is expected that a large or early release frequency below 10^{-6}/r.y could be achieved for events of internal origin. Lower values could also be considered as a safety target (for orienting the design process), but not as a limit. Some examples are listed below:

a) China:
   The value of 10^{-7}/r.y is recommended to be adopted as a complementary judgement for 'practical elimination' of individual sequences.
b) Finland:
   In order to limit long term effects, the probability of exceeding the limit of 100 TBq for atmospheric releases of Cs-137 should be lower than 5×10^{-7}/r.y;
c) Japan:
   The following numerical targets were used by Hitachi GE for the design of the ABWR:
   i.   Total large release frequency (LRF) of 10^{-6}/r.y;
   ii.   A value of 10^{-7}/r.y for individual release categories (value considered for the 'practical elimination' of individual accident scenarios).
d) Republic of Korea:
   i.   Limits prescribed by the nuclear regulatory body:
         - The upper limit on the effective exposure dose at the exclusion area boundary in case of a nuclear accident is 250 mSv [29];
         - The sum of accident frequencies that may lead to a release of the radionuclide Cs-137 over 100 TBq should not exceed 10^{-6}/r.y [30];
   ii.   Targets set by reactor designers:
         - For the design of the new reactor APR1000, the value of 10^{-7}/r.y is used as a quantitative target for the demonstration of 'practical elimination' of individual sequences through a demonstration of 'extreme unlikeliness' (this value is associated to results of level 3 PSA analyses, see the related subsection in Annex IV for additional details);

---

[21] Design measures for external hazards are discussed in Section°6.

— For the design of the new reactor iPower, the containment failure frequency should be less than 10^{-9}/r.y to 'practically eliminate' the radioactive material release through the containment.

The above values are defined for internal events only, excluding external hazards[22].

4.3.2. Examples of demonstrations of 'practical elimination'

Provisions that are typically used for the demonstration of 'practical elimination' of each of the conditions presented in Table 9 are summarized in Table 10.

TABLE 10. PROVISIONS TYPICALLY USED TO DEMONSTRATE 'PRACTICAL ELIMINATION'

Subcategory	Typical provisions / Elements of demonstration
Failure of a large component in the RCS	• Assignment of equipment to the highest safety class • Application of a high level of requirements for design, manufacture (highest level of construction codes), installation • Monitoring of acceptable defects and their development, using highest levels of requirements for in-service inspection codes • Operational rules for chemistry monitoring and control (including post-accident sampling system) • Operational rules to limit the variation of the reactor temperature in the acceptable range (e.g. avoid quick heating-up or cooling-down) • Limitation/management of vessel embrittlement due to neutron flux • Probabilistic fracture mechanical analysis
Uncontrolled reactivity accidents	• Core design ensuring a negative reactivity coefficient • Speed limits for control rods/clusters extraction movements • For PWRs, design solutions excluding or limiting interfacing leaks to primary circuit, automatic or manual isolation of sources of dilution and monitoring of boron concentration
High-pressure melt ejection (HPME) and direct containment heating (DCH)	• Highly reliable discharge devices including dedicated discharge line, to decrease the pressure below a threshold value (around 20 bar) • Means to secure the long term operation of discharge devices, such as long term batteries • Emergency procedures to depressurize the RCS • Limitation of open space between the reactor pit and the openings towards the containment dome • For PWRs, demonstration of absence of RCS failure by creep damage of the surge line and SG tubes in case of depressurisation failure
In-vessel steam explosions leading to early containment failure	• Core design preventing fast move of molten fuel debris to the RPV bottom • Presence of a large void fraction in the RPV bottom • Design of internal structures favouring energy absorption in the RPV bottom

---

[22] Indeed, for some external hazards, and depending on the specific site, it may not be practicable or even possible to define frequencies of occurrence for high levels of magnitude, because of poor input data or limited length of time of the data. The topic of design against external hazards is further developed in Section°6.

TABLE 10. PROVISIONS TYPICALLY USED TO DEMONSTRATE 'PRACTICAL ELIMINATION' (cont.)

Subcategory	Typical provisions / Elements of demonstration
Ex-vessel steam explosions leading to early containment failure	• Implementation of in-vessel melt retention (IVMR) strategy to exclude the RPV failure • Design of a dry reactor cavity (avoidance of presence of water before the RPV failure)
Combustion of gases endangering the containment integrity	• Use of PARs inside the containment • Verification that the installed capacity and recombiners position is effective in limiting the risk of gas combustion • Limitation of MCCI gases by preventing MCCI (see below)
Molten core concrete interaction (MCCI)	• Implementation (if any) of IVMR strategy to exclude the RPV failure • Implementation (if any) of an ex-vessel reactor cooling strategy using a core catcher or a large spreading area with sacrificial concrete layer (e.g. siliceous concrete)
Late containment failure (LCF) due to overpressure or overheating	• Diversified means of containment cooling (including passive systems) and external spraying of metallic containments • Filtered containment venting
Severe accident with containment bypass at power	• Reliable containment isolation system (CIS) • Operational rules and operating procedures to limit the occurrence of a bypass
Core melt at shutdown state with RPV open	• Large water quantity in reactor cavity providing grace time to boiling • Means to cool the reactor in shutdown states • Diverse means to inject water in reactor well and prevent boiling • Adequate power supply provided to face SBO conditions • Procedures to isolate the containment and particularly the containment hatch very quickly (typically less than 1 hour)
Significant fuel degradation in the SFP by boil-off	• Geometrical provisions to avoid draining and siphoning of water and ensuring that fuel elements are covered in case of a leaking pipe connected to the SFP • Large water reserve providing make-up to avoid uncovering especially for long term passive (e.g. boiling) fuel cooling • Redundant and diverse cooling systems • Diverse make-up systems
Catastrophic failure of the SFP	• Design of SFP structure to withstand the effects of internal and external hazards • Provisions to exclude the possibility of heavy load drop that may threaten the SFP integrity

Generally, when the demonstration is based on a high level of confidence for the sequence to be extremely unlikely to arise, several provisions are combined with deterministic and probabilistic analyses to demonstrate the 'practical elimination'.

Some examples of such a demonstration are provided in Sections 4.3.2.1 and 4.3.2.2 below, while additional examples are provided in the Appendix.

4.3.2.1. 'Practical elimination' of high-pressure melt ejection

The demonstration of 'practical elimination' of HPME may be composed of one or more of the following:

- Evaluation of several scenarios to determine the bounding case for the discharge capacity. These scenarios may include loss of off-site power with unavailability of all diesel generators, failure of primary and secondary feed and bleed;
- Installation of a discharge system independent from the overpressure protection system. This system may be tolerant to single failure and is able to operate in case of total loss of AC power supply (batteries, fail-safe open valves, gas-operated valves, etc.). It needs to be qualified to accidental fluid conditions and ambient conditions;
- Limitation of openings between the reactor pit and the containment dome to limit heating of the containment atmosphere by the corium;
- Evaluation of the behaviour of RCS components submitted to high temperature gases to demonstrate that, even in case of failure of the discharge system, a pipe will fail by creep damage prior to the RPV failure;
- Evaluation of the probability of HPME to demonstrate that this is extremely low.

More detailed examples of 'practical elimination' of HPME and DCH on various reactor designs are provided in the Appendix.

4.3.2.2. 'Practical elimination' of containment failure due to combustion of gases

Similarly, the demonstration of 'practical elimination' of a containment failure due to combustion of gases may rely on one or more of the following:

- Evaluation of several scenarios to determine the bounding cases including fast or slow rate of hydrogen production, hydrogen flow paths;
- Installation of several PARs at various places within the containment;
- Design of the containment to provide a large free volume and hydrogen flow paths to facilitate dispersion and mixing of hydrogen;
- Implementation of either IVMR or ex-vessel reactor cooling (e.g. design of a core catcher) to minimize the production of combustible gases outside the RPV;
- Computational fluid dynamic calculations (including sensitivity analyses) to check that high concentrations of combustible gases are sufficiently limited for all scenarios.

Recent international research publications provide more information on hydrogen management, and on computer codes that may be used for the safety demonstration [31]. Distribution of hydrogen in containment systems was specifically investigated in Ref. [32], using experimental tests and benchmark calculations.

5. DEFINING SAFETY MARGINS TO AVOID CLIFF EDGE EFFECTS

Subsection 5.1 introduces relevant IAEA safety standards while subsection 5.2 illustrates current practices in the adoption of safety margins to avoid cliff edge effects.

5.1. RELEVANT IAEA SAFETY STANDARDS

The term 'cliff edge effect' is defined in the IAEA Safety Glossary [7] as follows:

"In a nuclear power plant or nuclear fuel cycle facility, a cliff edge effect is an instance of severely abnormal facility behaviour caused by an abrupt transition from one facility status to another following a small deviation in a facility parameter, and thus a sudden large variation in facility conditions in response to a small variation in an input."

In SSR-2/1 (Rev. 1) [1], as part of the lessons learned from the Fukushima Daiichi accident, emphasis has been put on requiring safety margins to avoid cliff edge effects. This is also reflected in SSG-2 (Rev. 1) [13], which provides recommended ways to fulfil requirements of SSR-2/1 (Rev. 1) [1] and GSR Part 4 (Rev. 1) [3] pertaining to deterministic safety analysis.

5.1.1. Requirements from SSR-2/1 (Rev. 1)

Requirements for the design of an NPP to ensure that small deviations in plant parameters do not lead to a cliff edge effect and that adequate margins are available to avoid cliff edge effects are summarized hereafter:

- SSR-2/1 (Rev. 1) [1] Section 4 (principal technical requirements), para. 4.11 (b)) states that "the design […] shall be conservative, and the construction shall be of high quality, so as to provide assurance that failures and deviations from normal operation are minimized, that accidents are prevented as far as is practicable and that a small deviation in a plant parameter does not lead to a cliff edge effect";
- SSR-2/1 (Rev. 1) [1] Section 5 (general plant design): Requirement 17 (internal and external hazards), Requirement 31 (ageing management), Requirement 42 (safety analysis of the plant design) para. 5.73 which states that "the safety analysis shall provide assurance that uncertainties have been given adequate consideration in the design of the plant and in particular that adequate margins are available to avoid cliff edge effects and early radioactive releases or large radioactive releases".

5.1.2. Recommendations and guidance from SSG-2 (Rev. 1) on how to comply with the safety requirements

SSG-2 (Rev. 1) [13] stresses in a variety of recommendations the necessity for the design of an NPP to foresee (as well as for the safety analysis to demonstrate with a high level of confidence) the existence of:

- Significant margins (para. 6.1);
- Adequate safety margins (paras 2.1, 7.1);
- Margins with respect to safety criteria (para. 2.6);
- Significant margins to the safety limits (para. 7.40);
- Margins to the loss of integrity of physical barriers (para. 4.12), sufficient margin between the criterion and the physical limit for loss of integrity of a barrier (para. 4.3);

- Safety margins applied to the design of SSCs, commensurate with the uncertainty in the loads they may have to bear, and with the consequences of their failure (para. 4.17).

In addition, paras 6.6, 6.7, 7.45 and 7.55 provide recommended ways to fulfil the requirements concerning the absence of cliff edge effects.

5.2. CURRENT PRACTICES IN THE ADOPTION OF SAFETY MARGIN TO AVOID CLIFF EDGE EFFECTS

Practices in the implementation of these requirements are presented in this section. Note that this section only deals with the adoption of margins to avoid cliff edge effects in the case of PIEs due to internal events and internal hazards. The consideration of margins to prevent cliff edge effects possibly originated by external hazards addressed further in Section 6.

5.2.1. Notions of cliff edge effects and safety margins

The concept of 'cliff edge effect' was extensively used in the aftermath of the accident at the Fukushima Daiichi NPP, for example in the European stress tests [33], [34], for which an underlying objective was to identify all the cliff edges and quantify the corresponding margins beyond the design basis. Examples of cliff edges given in Ref. [33] were "exceeding a point where significant flooding of plant area starts after water overtopping a protection dike or exhaustion of the capacity of the batteries in the event of a station blackout".

In the definition of the term 'cliff edge effect' provided in the IAEA Safety Glossary [7], the expression "an instance of severely abnormal facility behaviour" for an NPP is understood as follows:

- For a DBA, the escalation to a DEC or a severe accident;

- For a DEC without significant fuel degradation, this would indicate an unacceptable damage to the fuel;

- For a DEC with core melting, this would indicate an unacceptable damage to the containment or, more generally, the failure of the last barrier.

Paragraph 5.73 of SSR-2/1 (Rev. 1) [1] establishes a link between the avoidance of cliff edge effects, the existence of adequate margins and the adequate consideration of uncertainties.

The terms 'margin' and 'safety margin' are not defined in the IAEA Safety Glossary [7]. Instead, IAEA-TECDOC-1791 [9] defines 'safety margin' as "the difference or ratio in physical units between the limiting value of an assigned parameter the surpassing of which leads to the failure of a structure, system or component, and the actual value of that parameter in the plant".

Reference [35] provides more information on the relationship between margins and uncertainties, and different views on the notion of safety margin. In general, the margin as defined in IAEA-TECDOC-1791 [9] can be subdivided into two different parts:

(i) The difference or ratio between the onset of damage and a value named "authorized limit" in the IAEA Safety Glossary [7]. The authorized limit represents "a limit on a measurable quantity, established or formally accepted by a regulatory body".

(ii) The difference between the authorized limit and the actual value of the parameter in the plant. This margin is available to the licensee.

A summary of the main practices in the consideration of Paragraph 5.73 of SSR-2/1 (Rev. 1) [1] is given in Section 5.2.2 below.

5.2.2. Design extension conditions

5.2.2.1. *Design extension conditions without significant fuel degradation*

In this domain, the cliff edge to be avoided is core melt. For DECs without significant fuel degradation that are similar to DBAs, the acceptance criteria or limits that are used are generally the same (for instance in LOCA without low pressure safety injection, the same fuel temperature limit), while in other cases specific decoupling limits may be adopted to provide margins to cliff edge.

As already introduced in Section 2, analyses of DECs without significant fuel degradation do consider uncertainties, but with reduced penalties compared to DBAs and with less conservative rules (such as no additional single failure given the number of failures already considered in such scenarios), taking into account the low likelihood of DECs without significant fuel degradation. Consideration of uncertainties varies among different countries; for example for the Flamanville-3 (FA3) EPR in France, the dominant parameters are defined with a high level of confidence (typically 95%), whereas other countries perform sensitivity studies.

All these practices avoid cliff edge effects because the acceptance criteria already include margins to the threshold values of cliff edge effects. Therefore, if the results meet the acceptance criteria, a margin is ensured and, hence, reasonable assurance is provided that a small deviation in a plant parameter will not lead to a cliff edge effect. The explicit consideration of uncertainties (as in the DBA domain) leads to a higher margin to cliff edge.

In the design of equipment for DECs without significant fuel degradation, the loads are defined in a similar way as for DBAs but using a best estimate plus uncertainty approach for determining the accident scenario. Stress limits justifying the integrity or operability of equipment are generally the same as those used for DBAs.

5.2.2.2. *Design extension conditions with core melting*

In this domain, the cliff edge to be avoided is the containment failure, which would lead to large or early radioactive releases. For new NPPs, the design of the containment building considers loads that result from DECs with core melting. Design provisions are implemented to prevent overpressurization of the containment, stabilize the molten core, remove the heat from the containment, and prevent gas combustion regimes from challenging the containment integrity.

SSG-53 [14] provides recommendations regarding the definition of acceptance criteria. It also recommends that failure modes and ultimate capabilities be evaluated.

Practices in the analysis of DECs with core melting vary among countries. Assumptions are generally less conservative than for DBAs, using deterministic best estimate studies, bounding the physical phenomena that may occur in a low-pressure core melt scenario (as high-pressure sequences are 'practically eliminated', see Section 4). They generally do not consider the SFC.

However, even if the general methodology is best estimate, some assumptions may be bounding values, for instance the hydrogen production due to fuel coolant interaction is maximized. Some countries, such as Finland, consider an additional single failure in their analysis.

Additional sensitivity calculations are provided to demonstrate the robustness of the safety features for DECs with core melting and verify that there are no cliff edge effects in terms of the loss of the confinement function. In China, for example, for the design of CNNC's HPR1000, sensitivity studies were performed to quantify the impact of a delayed time of initiation of RCS dedicated depressurization valves, with respect to the setting point established for a core outlet temperature of 650°C. Calculation results have confirmed the effectiveness of the depressurization safety function even in the presence of a delayed time of initiation.

Uncertainties for some of the phenomena encountered during severe accidents may be extremely high. They generally follow a decreasing trend with time due to the building of additional knowledge, in part due to international R&D programmes. References [36], [37], [38] provide the state of the art on some of these phenomena (ex-vessel steam explosion, MCCI, ex-vessel molten core coolability, hydrogen management), and help to identify the relevant parameters to be studied in the frame of the sensitivity studies. Other sources may also be used for this purpose, for example:

- NUREG series [39], [40], [41], [42];
- Electric Power Research Institute (EPRI) utility requirements for LWR [43];
- Modular Accident Analysis Program, Version 5 (MAAP5) applications guidance [44].

Where uncertainties are very important, appropriate design choices may enable the elimination or reduction of their impact; for instance, having a dry reactor pit reduces the risk of ex-vessel steam explosion in case of RPV failure.

Below are two sets of detailed practices from recent projects.

(a)   ABWR (Japan)

Severe accident safety analyses enable the evaluation of the grace time for accident management that is then reflected into the level 1 and level 2 PSA.

These are also used to evaluate the magnitude and characteristics of the source term used for level 3 PSA.

The conclusion of sensitivity analyses conducted on the ABWR shows that the impact of the parameters below is not significant:

- Impact of the time margin (accident progression): the results of the evaluation conclude that timing of key events does not significantly impact the accident sequence evolution.
- Impact of the source term: the results of the evaluation conclude that the amount of fission product release is not substantially changed, however an increase in some fission product groups is observed. A conservative value of the failure surface in the containment is assumed in the base case of the severe accident analysis and the amount of fission product release decreases substantially if a more realistic failure surface is assumed. Therefore, the impact on the risk due to the uncertainty of model parameters is smaller than the one due to the conservative failure surface.

(b)   APR1000 (Republic of Korea)

In order to bound the severe accident sequences, a set of deterministic analyses using conservative assumptions is performed to envelop the core melt progression and its consequence (i.e. containment failure). In addition, uncertainty analyses and/or sensitivity analyses are performed to incorporate the uncertainties in the core melt accidents. The uncertainty analyses include Latin hypercube sampling for highly contributing parameters, and the sensitivity analyses select conservative parameter values within the range of reasonable physical or numerical models. The results of uncertainty or sensitivity analyses are used to demonstrate that the containment integrity is maintained with a high level of confidence such that cliff edge effects are avoided.

6. DESIGN FOR EXTERNAL HAZARDS

Subsection 6.1 introduces relevant IAEA safety standards while subsection 6.2 illustrates the experience in the implementation of requirements from SSR-2/1 (Rev. 1) [1] for the design and safety demonstration of new NPPs.

6.1. RELEVANT IAEA SAFETY STANDARDS

Guidance from SSR-2/1 (Rev. 1) [1] and related safety guides is introduced in the following subsections.

6.1.1. Requirements from SSR-2/1 (Rev. 1)

Requirements related to the consideration of external hazards in the design of NPPs are summarized hereafter:

- SSR-2/1 (Rev. 1) [1] Section 5 (general plant design): Requirement 17 (internal and external hazards), deals specifically with external hazards:
    - Para. 5.17: "[…] In the short term, the safety of the plant shall not be permitted to be dependent on the availability of off-site services such as electricity supply and firefighting services. The design shall take due account of site-specific conditions to determine the maximum delay time by which off-site services need to be available".
    - Para. 5.19: "Features shall be provided to minimize any interactions between buildings containing items important to safety (including power cabling and control cabling) and any other plant structure as a result of external events considered in the design".
    - Para. 5.21: "The design of the plant shall provide for an adequate margin to protect items important to safety against levels of external hazards to be considered for design, derived from the hazard evaluation for the site, and to avoid cliff edge effects".
    - Para. 5.21A: "The design of the plant shall also provide for an adequate margin to protect items ultimately necessary to prevent an early radioactive release or a large radioactive release in the event of levels of natural hazards exceeding those considered for design, derived from the hazard evaluation for the site".
- SSR-2/1 (Rev. 1) [1] Section 6 (design of specific plant systems): Requirement 65 (control room), para. 6.40A states that "the design of the control room shall provide an adequate margin against levels of natural hazards more severe than those considered for design, derived from the hazard evaluation for the site".

6.1.2. Recommendations and guidance from IAEA Safety Guides on how to comply with the safety requirements

Recommended ways to fulfil the abovementioned requirements are illustrated in the following IAEA Safety Guides:

- IAEA Safety Standards Series No. SSG-68, Design of Nuclear Installations Against External Events Excluding Earthquakes [45];
- IAEA Safety Standards Series No. SSG-67, Seismic Design for Nuclear Installations [46].

6.2. EXPERIENCE IN THE IMPLEMENTATION OF SSR-2/1 (REV. 1) REQUIREMENTS

In this section, the following terminology will be adopted:

i. Terminology from the IAEA Safety Glossary [7]:
   - External events: events unconnected with the operation of a facility or the conduct of an activity that could have an effect on the safety of the facility or activity.
   - Design basis external event: the external event(s), or combination(s) of external events, considered in the design basis of all or any part of a facility[23].
ii. Terminology consistent with para. 5.17 of SSR-2/1 (Rev. 1) [1]:
   - External hazard: external event either of natural origin or human induced that has been identified in the site evaluation process.

The consideration of external hazards in the design of an NPP typically foresees the following steps:

- The identification of all the external events and their combinations to be considered in the design, as task of the site evaluation (requirements are established in IAEA Safety Standards Series No. SSR-1, Site Evaluation for Nuclear Installations [5]);
- The development of hazard curves for all selected events;
- The definition of a level of magnitude for each of the hazards to be considered in the design (the design basis);
- The list of structures and equipment to be specifically designed or protected against such hazards (see Refs [45], [46]);
- The definition of methodology, rules and assumptions for the safety assessment of their consequences on the plant (requirements are established in SSR-2/1 (Rev. 1) [1]).

The practical experience by contributors to this TECDOC in the consideration of design basis external hazards (Section 6.2.1) and beyond design basis external hazards (Section 6.2.2) is summarized hereafter.

6.2.1. Consideration of design basis external hazards

The following subsections illustrate practices in the identification of design basis external hazards (subsection 6.2.1.1), in the definition of the level of magnitude (subsection 6.2.1.2) and of the list of structures and equipment to be specifically designed or at least protected against design basis external hazards (subsection 6.2.1.3), and in the methodology applied for such external hazards (subsection 6.2.1.4).

6.2.1.1. Identification

External hazards are site specific, and various IAEA safety standards [5], [45], [46] identify typical hazards to be considered as design basis external hazards, applicable to the specific site and facility. Therefore, this topic will not be developed further in this TECDOC, which focuses on collecting information in the consideration of external events in the design of NPPs.

---

[23] The concept of 'beyond design basis external event' is not defined in the IAEA Safety Glossary [7] while it is defined in SSG-68 [45] "to indicate a level of external hazard exceeding those considered for design, derived from the hazard evaluation for the site". With this meaning the term is used throughout the remainder of this publication.

Each external hazard shows intrinsic specificities that need to be specifically considered: for example some hazards may relate to a determined part of the plant (e.g. the UHS for clogging hazard from sea or raw river water; see also Section 7), while others (e.g. earthquakes) affect the whole plant.

Requirement 1 of SSR-1 [5] states:

"The safety objective in site evaluation for nuclear installations shall be to characterize the natural and human induced external hazards that might affect the safety of the nuclear installation, in order to provide adequate input for demonstration of protection of people and the environment from harmful effects of ionizing radiation".

During site evaluation, Requirement 1 of SSR-1 [5] is addressed through identification of a list of external events of:

- Natural origin (e.g. earthquakes, floods, extreme meteorological conditions, extreme winds, dust and sandstorms, lightning, volcanism);
- Human induced origin (e.g. accidental aircraft crashes, explosion, fire, vehicle impact, release of corrosive and/or hazardous gases or liquids).

Depending on site characteristics, external events from the above list, as well as relevant combinations of events (e.g. high sea level and rain contributing to flood), where appropriate, are considered as design basis external events.

For example, in China, the identification of relevant combinations considers:

  i.   Combinations between hazards resulting from a common initiating phenomenon (e.g. precipitation and flood/extreme wind/lighting, extreme wind and flood);
  ii.   Combinations between an external event and its consequential effects, for example:
        - Earthquake and internal fire/flood/pipe failure/heavy loads;
        - Earthquake and external flood/explosion;
        - Loss of off-site power (LOOP) resulting from earthquake, flood, extreme wind, tornado, lighting, explosion, or aircraft crash;
        - Aircraft crash and explosion.
  iii. Combinations between an external event and an independent external event or postulated initiating event identified either by probabilistic analysis or considered for the assessment of design margins based on engineering judgement (e.g. earthquake and design basic accident). However, external hazards and PIEs that may occur independently (such as human induced external hazards and equipment failures or operator errors) are not considered as relevant combinations, unless such a combination is shown to have a sufficiently high probability of occurrence.

Similar types of combinations are considered in other countries, including France and Japan (Hitachi GE).

In the Russian Federation, for any independent external event considered for the design, its relevant combinations and interactions with other external event(s) should also be considered according to Ref. [47]. A logic flow chart of external event safety analysis (also including the results of the external hazards PSA) should be developed according to Ref. [48], where a detailed matrix of external hazard combinations from three natural elements (atmosphere, hydrosphere, and ground including fire) is considered.

6.2.1.2.   Definition of the level of magnitude of design basis external events

From the relevant data gathered during the site evaluation, for each considered external event (or combination of events), a level of magnitude is defined as an application of regulatory prescription or guidance and/or designer rules and guidance.

In China, the application of the Chinese national regulatory guide [49] and additional safety guides on external hazards [50], [51] combines deterministic and probabilistic methods applied to the site data to define the appropriate level of magnitude for considered events.

In the Republic of Korea, the approach is to derive, from the analysis of the site data, a level of magnitude bounding the maximum probable event, likely to occur over the plant lifetime.

In Japan, under the NRA Ordinance on Standards for Installation Permit [52], the regulatory requirements for external events (for both natural phenomena and human induced events) were significantly reinforced, benefitting from the lessons learned from the Fukushima Daiichi accident (for additional details on requirements related to earthquakes and tsunamis, see Annex III).

In the Russian Federation, the minimal design requirements set by the regulatory body are:

- For earthquakes, the minimum magnitude should be not less than $0.1g$ free field horizontal peak ground acceleration;
- External blast wave load to SSCs should be not less than 10 kPa with a compression phase duration up to 1 second;
- Aircraft and other external missiles applied to localizing safety structures should have a contact zone impact equivalent to a 5 tonne light aircraft.

In accordance with Ref. [47], site conditions and site evaluation data are required to estimate probabilistic levels of the design external events magnitude having 10^{-4} annual exceedance probability for natural external events (including earthquakes) and 10^{-6} annual exceedance probability for human induced external events (including up to 20 tonne weight aircraft crash).

In some countries (France, United Kingdom), as an application of regulatory guidance such as the Western European Nuclear Regulators' Association (WENRA) safety reference levels for existing reactors [26], a target value in terms of annual exceedance frequency is used[24] to characterize the level of magnitude of natural external events. When such frequency cannot be calculated with an acceptable degree of certainty, alternative approaches are provided, for example a higher frequency is considered, but an appropriate margin is added to characterize the level of magnitude corresponding to lower frequencies.

A good practice to facilitate the demonstration of the avoidance of cliff edge effects (para. 5.21 of SSR-2/1 (Rev. 1) [1]) and to account for future potential increases of the hazard magnitude

---

[24] The issue T (natural hazards) T4.2 in Ref. [26] (definition of the design basis events) foresees that "the exceedance frequencies of design basis events shall be low enough to ensure a high degree of protection with respect to external hazards. A common target value of frequency, not higher than 10^{-4} per annum, shall be used for each design basis event. Where it is not possible to calculate these frequencies with an acceptable degree of certainty, an event shall be chosen and justified to reach an equivalent level of safety".

in the lifetime of the facility (e.g. due to climate change), is to define such a level of magnitude in a conservative manner. Cliff edge effects are further discussed in Section 6.2.2, where a higher magnitude of external hazard levels is considered for beyond design basis external hazards.

Among the abovementioned approaches, common approaches are observed for the definition of the level of magnitude of various external hazards. Determination of the level of magnitude considers all available data. This includes historical data about past events at the site and in its vicinity, as well as data relevant to the hazard evaluation. For instance, in the case of earthquake, historical and measured geological data are used for the hazard analysis. Hazard curves are developed based on engineering evaluations of all models and findings. The treatment of available data can include deterministic and probabilistic considerations, with due consideration of uncertainties. When relevant, projected increases of the hazard magnitude level are addressed from the very initial design phases, to reduce the risks of large scale modifications in future periodical safety reviews. This is particularly the case for flood and weather related hazards, subjected to climate change. Finally, when the estimated frequency of occurrence of an external hazard (or a combination of external hazards) is sufficiently low, the hazard may either be excluded from further consideration or be submitted to a special treatment, as it is the case in China, France, Japan (Hitachi GE), Republic of Korea[25] and Russian Federation.

In some countries (e.g. France, Japan (Hitachi GE), Republic of Korea, Russian Federation), a given design may be developed or selected for several different sites. In that case, a 'generic' or 'standard' design is defined for the main plant SSCs (e.g. reactor buildings, safety systems buildings). Therefore, the list of hazards to be considered and their level of magnitude are defined such as to provide a bounding case meaning that the generic plant could be implemented at any of the considered sites, with only limited adaptation to account for site specificities.

6.2.1.3. *List of structures and equipment to be specifically designed or at least protected against design basis external hazards*

For the identified list of events and for their level of magnitude (as described in previous subsections), in accordance with Requirement 17 of SSR-2/1 (Rev. 1) [1]:

> "Items important to safety shall be designed and located, with due consideration of other implications for safety, to withstand the effects of hazards or to be protected, in accordance with their importance to safety, against hazards and against common cause failure mechanisms generated by hazard" (para. 5.15A);

> "The design of the plant shall provide for an adequate margin to protect items important to safety against levels of external hazards to be considered for design, derived from the hazard evaluation for the site, and to avoid cliff edge effects" (para. 5.21).

The aim of the protection of items important to safety against an external hazard is to prevent consequences on the plant, as far as achievable, and to ensure that despite the consequences on the plant, the remaining safety systems and DEC safety features will fulfil the safety functions required in such an event. The ultimate objective is to ensure that, if an external hazard were to

---

[25] For example, in the Republic of Korea, for the APR1000, if an accidental aircraft hazard has a lower probability than a chosen screening value, this hazard will not be considered within the design basis.

occur, radiation doses to workers at the plant and to members of the public do not exceed the dose limits (Requirement 5 of SSR-2/1 (Rev. 1) [1]).

To avoid entering into deep analysis to detail all possible sequences of hazards occurring on the plant, in several countries (like China, France, Republic of Korea, and Russian Federation), for most of the external hazards the following decoupling approach is adopted. The list of components to be designed or protected against an external hazard may include, within the list of items important to safety[26]:

- The safety systems and those potentially interacting with the safety systems as result of an external event;
- The safety features for DECs;
- A selected list of safety related items (e.g. the reactor, the RCS).

For example, in China, France and Japan (Hitachi GE), the whole set of safety systems required to mitigate DBAs is generally designed to withstand the effects of earthquakes.

In the perspective of the above aim, the items important to safety to be specifically designed or protected[27] against design basis external events may include the following:

- Equipment whose failure would lead to an accident;
- Structures housing items important to safety;
- The whole or some of the items part of safety systems and safety features necessary to manage DBAs, and DECs without significant fuel degradation (fulfil the safety functions required to reach a safe state);
- Safety features for DECs with core melting;
- Equipment specifically provided for protection against other external hazards, where relevant[28].

However, as explained in Section 6.2.1.1, some external hazards have intrinsic specificities, and therefore:

- For redundant equipment and for some hazards, a refined approach may be considered when such a systematic protection might not be reasonably practicable. In that case, the list of SSCs to be designed or protected against an external hazard is reduced.
- If the external hazard consequences may be limited to an area of the plant (e.g. aircraft crash, explosion), the list of components to be designed or protected may be limited to those in this area or limited to a list of redundant equipment physically separated from this area.
- For some external hazards (e.g. clogging, frazil ice) only affecting a specific part of the

---

[26] The IAEA Safety Glossary [7] defines the 'items important to safety' as being composed of three categories of SSCs:
  1. Safety systems (which are composed of: (i) protection system; (ii) safety actuation system; (iii) safety system support features);
  2. Safety features for design extension conditions;
  3. Safety related items.

[27] For some external hazards, components are protected by their housing building and do not require a specific design, whereas for components outside buildings a design specific to the hazard is required. For example, chimneys should be protected against high wind, whereas sensors and valves inside the containment do not need additional protection against high wind than the one provided by the containment itself.

[28] In Japan, seawalls for flood protection against tsunamis shall be of the S class (the highest class in the anti-seismic design classification), so that the flooding prevention functions are not lost due to earthquakes.

plant, the list of components to be designed or protected against an external hazard may be reduced to some equipment in this part of the plant or physically separated from this part: for example the filtering system only, a remote alternate UHS, a remote diesel or a specific protection equipment.

6.2.1.4. *Methodology, rules and assumptions*

Practices might be summarized as follows.

(a) Design process

Possible approaches for the design stage include the following:

a) SSCs design derived from the design approach for civil buildings: the approach of the designer (in China, France, Japan (Hitachi GE) and Russian Federation) to protect a NPP against design basis external events primarily commences at an early stage of the design and is derived from codes and standards for the design of buildings. Indeed, even for non-nuclear applications, the first aim of a building is to provide a shelter and a protection against at least weather conditions, hence standards (such as the European EN Eurocode[29] [53]) consider 'load cases' for the protection against natural hazards such as rain, snow, fire, wind, and earthquake.

In China, France, Japan (Hitachi GE), Republic of Korea, and Russian Federation, such an approach is applied for the earthquake hazard, at least for the safety systems and the buildings housing them. Practically, suppliers design SSCs accounting for such a defined load case. The aim is to either:
- Prevent the consequences of external hazards;
- Prevent harmful consequences on the safety functions required for AOOs, DBAs or DECs. This means not adversely affecting the items important to safety fulfilling the safety functions required to face the external hazard or at least avoiding CCF on redundant equipment performing such safety functions;
- Prevent the consequences on equipment whose failure may initiate another internal hazard.

b) Hazard assessment of the plant: where it is not possible to fully apply the previous approach or as part of the hazard assessment of the plant, detailed hazard studies are provided to evaluate the effects and the consequences of the external hazards to be considered for their level of magnitude as defined above.

c) Practically, the following 'hybrid approach' may be adopted:
- Design the plant considering the full range of PIEs for different plant states (operational states and accident conditions).
- Design the structures and components, required for these conditions, against some of the external hazards, where the abovementioned approach is practicable. Examples of such provisions are an external shield for aircraft crash, or the building base isolation for earthquake. This implies the application of relevant national and international codes and standards, as highlighted in Requirement 9 of SSR-2/1 (Rev. 1) [1].
- Apply layout rules such as physical separation and other appropriate rules to avoid

---

[29] A series of ten European Standards, providing a common approach for the design of buildings and other civil engineering works and construction products.

or limit the consequences of external (and internal) hazards, which may include the provision of dedicated passive and active means. Examples of such provisions are: 'dry site' for flood; site protection dam for flood; barriers for explosions.

- Assess the plant against external (and internal) hazards, as detailed hereafter.

(b)   Safety assessment

In parallel to the design process (described above), the safety assessment[30] aims to demonstrate that the risk posed by external events is below acceptable limits, meaning that sufficient SSCs are available to reach and maintain a safe state, as well as to limit the radiological consequences under the acceptable limits defined with reference to Requirement 5 of SSR-2/1 (Rev. 1) [1]. The safety assessment is also used to calculate the margin for beyond design basis events and to exclude the possibility of occurrence of cliff edge effects.

The assessment generally credits all the measures in place to limit the consequences:

- As part of load cases considered in SSC design;
- In terms of layout rules and prevention means already designed.

The assessment may consider the predictability and the characteristics of the external hazard to forecast actions[31], prior to the event reaching the site (e.g. the closure of all openings in case of high wind, tornado, or cold weather) or to trigger the associated alert systems (fire), when relevant.

The assessment is an iterative process conducted in parallel with the design and may lead to identify the need for additional protection measures where necessary, especially at early design stages. The final solution is expected to represent the best balance among safety aspects, operational aspects and other factors.

Finally, to provide additional DiD to the basic forms of protection defined above, for some external hazards, proactive, active, or administrative measures based on forewarning can also provide safety benefits. The reliability ascribed to such measures has to be commensurate with the reliability of the monitoring and forecasting equipment and operator reliabilities.

(c)   Rules and methods

Generally, an assessment methodology or guidance from the regulator (China, Japan) and/or the designer (France, Republic of Korea, Russian Federation) specific to each external hazard provides a structured approach to be followed.

All equipment that may fail due to the external hazard is considered as failed (China, France, Japan (Hitachi GE), Republic of Korea, Russian Federation) and a single active failure may be considered on the active hazard protective measures (France, Russian Federation). Where an event like an AOO or a DBA cannot be avoided, the rules and acceptance criteria applicable to AOOs or DBAs are followed, but the safety analyses do not credit the equipment failed due to the hazard (note that the failure of a support system implies the failure of the supported components, unless backed-up).

---

[30] See for example Requirement 10 of SSR-2/1 (Rev. 1) [1] and, more generally, the requirements from GSR Part 4 (Rev. 1) [3].
[31] On the basis of monitoring means or forecast means (weather).

For earthquakes, this deterministic assessment is complemented by PSA (Japan (Hitachi GE), Republic of Korea, Russian Federation).

6.2.2. Consideration of beyond design basis external hazards

As explained in Footnote 22 (see Section 6.2), beyond design basis external events are those involving a level of natural hazards exceeding those considered for design, derived from the hazard evaluation for the site.

The consideration of these events, as a complement to design basis external events, was introduced within SSR-2/1 (Rev. 1) [1] in some requirements, notably para. 5.21A:

> "The design of the plant shall also provide for an adequate margin to protect items ultimately necessary to prevent an early radioactive release or a large radioactive release in the event of levels of natural hazards exceeding those considered for design, derived from the hazard evaluation for the site".

In the Republic of Korea and Japan (Hitachi GE), those beyond design basis external events are only due to natural external hazards. However, in other countries, beyond design basis external events include some human induced external hazards (even though para. 5.21A requires specifically the consideration of natural hazards only), for example:

- In the Russian Federation, for the design of the WWER-TOI, an accidental military aircraft crash (mass 20 tonnes, impact velocity of 215 m/s) is considered as a design basis external event, while an accidental large commercial aircraft crash (mass 400 tonnes, impact velocity 200 m/s) is considered as a beyond design basis external hazard.
- In China, a beyond design basis aircraft crash event was considered in the design of CNNC's HPR1000.

6.2.2.1. Identification

As part of beyond design basis external events, earthquakes and potential external flooding (and their consequential effects, such as tsunamis or LOOP) are generally considered.

Furthermore, in France, a tornado event, lightning, and heavy rainfall associated with an extreme flooding event are specifically considered. For the next generation of EPRs, other natural events (weather related hazards) may also be specifically considered where relevant.

In the Republic of Korea, the rare and severe external hazard (RSEH), as indicated in the European Utility Requirements Rev. E [54], is considered. The design for RSEHs aims to verify that the final overall probabilistic evaluation using a realistic approach and best estimate rules meets the objectives on core melt and radioactive releases. The NPPs are also designed to cover large uncertainties which could exist for hazards, by ensuring that sufficient margin exists regarding cliff edge effects.

In countries where a safety margin assessment approach is followed, the entire spectrum of design basis external hazards may be considered.

6.2.2.2. Definition of the level of magnitude of beyond design basis external events

For the specifically considered events (at least for earthquakes), higher levels of magnitude with respect to design basis external events are derived with various approaches:

- Consideration of a lower frequency of occurrence (China[32], Republic of Korea) for such events exceeding those considered for the design basis;
- By using a multiplicative coefficient (e.g. 1.5 for seismic levels, whereas for locations with higher seismic levels a more reasonable value is considered) and/or a fixed reasonable margin (in the Republic of Korea[33] and Russian Federation).

6.2.2.3. *List of structures and equipment to be specifically designed or at least protected against beyond design basis external hazards*

In accordance with para. 5.21A of SSR-2/1 (Rev. 1) [1], the aim of the protection of items important to safety against a beyond design basis external hazard is to ensure that despite the consequences on the plant, a sufficient number of safety systems and DEC safety features remain operational to reach and maintain a safe state and ultimately to limit the radiological consequences with the aim to prevent an early or a large radioactive release. Considering the unlikelihood of such hazards beyond the design basis, this may refer to a limited number of equipment ultimately necessary to prevent an early or a large radioactive release.

The identification of SSCs requiring protection from beyond design basis external natural hazards includes:

- Features for DECs with core melting;
- Equipment supporting the demonstration of 'practical elimination' of events that could lead to an early radioactive release or a large radioactive release (through the approach detailed in Section 4);
- The equipment necessary or allowing the use of NPE, i.e. the features mentioned in Requirements 58, 68 and 80 of SSR-2/1 (Rev. 1) [1] as being necessary to connect NPE (for example the connecting points allowing to inject water, provide power supply and extract heat from the containment, see Section 8 for additional details).

In accordance with Requirement 65 of SSR-2/1 (Rev. 1) [1], the list may also include the main control room or an alternative one, to monitor and control the equipment, and to ensure adequate communications.

6.2.2.4. *Methodology, rules and assumptions*

Practices related to the overall approach, as well as rules and methods, might be summarized as follows.

(a) Approach

In order to fulfil para. 5.21A of SSR-2/1 (Rev. 1) [1], two different approaches may be followed by designers, depending on the type of hazard:

   a) Define load conditions for the beyond design basis external events and consider them in the design of the structures and equipment (those in the list of items identified in

---

[32] For the design of CNNC's HPR1000, based on the lessons learned from the Fukushima Daiichi accident, the beyond design basis for flood is set by a combination of the design basis flood and the historical record of precipitation on a return period of 1000 years.

[33] For the design of the lightning protection system for the APR1000, an intensity of 200 kA is considered for the design basis hazard and 300 kA for the RSEH.

Sections 6.2.2.1–6.2.2.3 as being ultimately necessary), with a similar approach to the one described in Section 6.2.1.4 for design basis external events;

b) Assess the margins of the plant against the level of magnitude of the beyond design basis external events considered.

Indeed, the two approaches are not excluding themselves: for some countries the first approach may constitute a preparatory step for the second one, with the ultimate objective of demonstrating the achievement of an appropriate level of safety for the plant:

a) The first approach is typically followed as part of the early stages of the design of structures for new designs (in China, France, Japan and Russian Federation), to reduce the risk to discover insufficient margins at a later stage.
For example, in China, a beyond design basis aircraft crash event is considered in the design of CNNC's HPR1000 through a strengthened design for the containment structure, fuel building and electrical building;

b) As part of the second approach, margins can be credited to demonstrate the avoidance of cliff edge effects and the capability of the plant to withstand hazard levels more severe than those considered in the design basis:

- Indeed, the design of SSCs to withstand a specific hazard follows relevant codes and standards. Such design codes and standards, as well as the approaches followed to protect the plant against design basis external hazards, include margins that enable the plant structure and equipment to withstand higher loads.

- In terms of earthquake, the application of a seismic margin assessment (in Japan and the Republic of Korea) or a full seismic PSA (in China, France[34] and Russian Federation) allows to assess the margin through an evaluation of the plant risks in terms of the core damage frequency (CDF) or potential for radioactive releases.

- In the Republic of Korea, for the design of the APR1000, a PSA-based seismic margin assessment is used, in accordance with recommendations of SECY-93-087 and ISG-020, which satisfies the European Utility Requirements Rev. E [54]. The PSA-based seismic margin assessment allows identifying potential vulnerabilities and assessing seismic margins for beyond design basis hazards.

- In the Republic of Korea, for the design of the APR1000, for the extreme air temperature hazard considered as RSEHs, SSCs used in DECs without significant fuel degradation initiated by RSEHs are designed considering a design basis external hazard only; however, these SSCs are eventually verified against RSEH air temperatures to assess whether the safety objectives (i.e. to prevent large or early radioactive releases and to require only limited protective measures) are fulfilled.

In addition, as part of mitigating measures for accident management, measures that facilitate the use of NPE for power supply and cooling after a beyond design basis external hazard are generally incorporated in NPPs (this topic is further discussed in Section 8).

(b)   Rules and methods

The overall methodology, assumptions and acceptance criteria are associated to performance requirements for SSCs that differ from those for design basis external events.

---

[34] In addition to the deterministic analysis.

Whether in the load conditions approach, in the margin assessment or in the supporting PSA studies, the basis (in China, Republic of Korea, and Russian Federation) is to follow a best estimate approach.

In France, levels of natural hazards exceeding those considered for design, derived from the hazard evaluation for the site, are considered in the design basis of the plant to ensure the application of proven technologies practices and well-established safety rules to the involved SSCs and associated safety demonstration. This approach is mainly related to natural hazards but is also applied, to some extent, to human induced hazards (depending on the context). The approach is different between currently operating reactors and new reactors. For a new reactor design, it is based mainly, but not exclusively, on the development of the concept of 'design extension hazards'.

The consequential effects of the beyond design basis external hazard are considered in terms of other hazards (e.g. flood from pipe rupture) or in terms of potential consequential LOOP.

7. STRENGTHENING THE CAPABILITIES FOR THE HEAT TRANSFER TO AN ULTIMATE HEAT SINK

Following the accident at the Fukushima Daiichi Nuclear Power Station in March 2011, an extensive and detailed assessment of NPP vulnerabilities to external hazards (including their combinations), such as flooding and earthquake, was initiated as part of the IAEA Action Plan on Nuclear Safety. In the case of the Fukushima Daiichi accident, a station blackout and the unavailability of emergency diesel power supply disrupted fuel cooling. Within this context, strengthening capabilities for the heat transfer to the UHS became a topic of interest.

The IAEA Safety Glossary [7] defines the UHS as "a medium into which the transferred residual heat can always be accepted, even if all other means of removing the heat have been lost or are insufficient. This medium is normally a body of water or the atmosphere".

Practically — the water or the air being the UHS — these media are never lost, while the access to them through SSCs may be lost. The so-called loss of the UHS event was initially defined as the loss of the access to the heat sink for plant with a single access to the heat sink. For new NPP designs, this is generally defined as the loss of the main UHS.

The objective of this section is to summarize current practices by Member States in strengthening the capabilities for heat transfer to the UHS in the design of new NPPs.

The section provides examples and practices incorporated in various NPP designs to strengthen the robustness of heat transfer pathways to the UHS.

7.1. REQUIREMENTS FOR THE HEAT TRANSFER TO THE ULTIMATE HEAT SINK

Guidance from SSR-2/1 (Rev. 1) [1] and national or international requirements are introduced in the following subsections.

7.1.1. Requirements from SSR-2/1 (Rev. 1)

Requirements dealing with the heat transfer to the UHS are summarized hereafter:

- SSR-2/1 (Rev. 1) [1] Section 6 (design of specific plant systems): Requirement 51 (removal of residual heat from the reactor core), Requirement 70 (heat transport systems), Requirement 80 (fuel handling and storage system), Requirement 53 (heat transfer to the UHS), where para. 6.19A requires that "systems for transferring heat shall have adequate reliability for the plant states in which they have to fulfil the heat transfer function. This may require the use of a different ultimate heat sink or different access to the ultimate heat sink".

7.1.2. Examples of specific national or international requirements

In the United States, the revised regulatory guide on UHS for NPPs was issued in 2015 [55]. The UHS requirement has been established to ensure provision of sufficient cooling for at least 30 days to enable safe shutdown and cool down of all nuclear reactor units. The guide also requires procedures for ensuring a continued capability of the UHS to provide sufficient cooling after 30 days.

In Canada, regulatory guides [56], [57] have specific requirements for the UHS:

- The regulatory document RD-367 [56] states that

  "the design shall include systems for transferring residual heat from SSCs important to safety to an ultimate heat sink. This function shall be highly reliable during normal operation, AOOs and DBAs. All systems that contribute to the transport of heat by conveying heat, providing power or supplying fluids to the heat transport systems shall be designed in accordance with the importance of their contribution to the function of heat transfer as a whole. Natural phenomena and human-induced events shall be considered in the design of heat transfer systems and in the choice of diversity and redundancy, both in the ultimate heat sinks and in the storage systems from which fluids for heat transfer are supplied. The design shall extend the capability to transfer residual heat from the core to an ultimate heat sink in the event of a severe accident".

- Similarly, the regulatory document REGDOC-2.5.2 [57] states that

  "the design shall include systems for transferring residual heat from SSCs important to safety to an ultimate heat sink. This overall function shall be subject to very high levels of reliability during operational states, DBAs and DECs. All systems that contribute to the transport of heat by conveying heat, providing power, or supplying fluids to the heat transport systems, shall be therefore designed in accordance with the importance of their contribution to the function of heat transfer as a whole. Natural phenomena and human induced events shall be considered in the design of heat transfer systems, and in the choice of diversity and redundancy, both in the ultimate heat sinks and in the storage systems from which fluids for heat transfer are supplied. The design shall extend the capability to transfer residual heat from the core to an ultimate heat sink so that, in the event of a severe accident considered as a DEC:

  1. Acceptable conditions can be maintained in SSCs needed for mitigation of severe accidents;
  2. Radioactive materials can be confined;
  3. Releases to the environment can be limited".

WENRA[35] has defined several reference levels pertaining to the heat removal function in issues E (design basis envelope) and F (design extension) of its safety reference levels for existing reactors [26]:

"Means for removing residual heat from the core after shutdown and from spent fuel storage, during and after anticipated operational occurrences and design basis accidents, shall be provided taking into account the assumptions of a single failure and the loss of off-site power" (issue E9.9).

"There shall be sufficient independent and diverse means including necessary power supplies available to remove the residual heat from the core and the spent

---

[35] WENRA members commit themselves to include these reference levels in their own national regulations.

fuel. At least one of these means shall be effective after events involving external hazards more severe than design basis events" (issue F4.7).

France has recently published a safety guide [58] containing requirements for the design of PWRs. It contains several recommendations about the design of the systems removing heat to the heat sink and more generally about the removal of the thermal heat produced by the nuclear fuel.

The architecture, the specified requirements and the reliability of the systems removing the heat produced by the fuel and dissipated by the various SSCs towards the heat sink have to be consistent with the architecture and the overall requirements defined for the SSCs that cool them.

Measures have to be taken to prevent risks of heat sink failure associated with external hazards. The need for specific measures, such as the distancing or diversification of water intakes, or the constitution of an emergency reserve has to be assessed on the basis of a characterization study of the site and an assessment of the vulnerability of the main heat sink.

In order to place multiple barriers between the systems carrying radioactive fluid — in particular the primary coolant — and the environment, usual practice for most PWRs in the design of the systems carrying heat to the heat sink is to include an intermediate cooling system between the heat exchangers cooling the systems carrying the radioactive fluid and the systems carrying the raw water.

7.2. EXPERIENCE IN THE IMPLEMENTATION OF SSR-2/1 (REV. 1) REQUIREMENTS

Subsection 7.2.1 introduces specific features depending on the nature of the heat sink, while subsection 7.2.2 illustrates practices in providing diverse access to the UHS and subsection 7.2.3 introduces practices in the demonstration of protection against levels of natural hazards exceeding those considered for design.

7.2.1. Different types of heat sink

As mentioned in the introduction, the UHS may be a body of water or the atmosphere (air cooling), with different possibilities offered in each case.

(a)   Water as UHS

Depending on the site, water may be provided from the sea, a river, or a lake/pool, and hence potentially subjected to external hazards (such as clogging, icing of the plant water intake filters, drought, pump flooding). For a water heat sink, diversity may be provided by several means, for instance:

- Creating an artificial lake independent from the river/sea, with a water intake on the lake and a second one on the river/sea. The lake enables reduction of some risks, as some parameters (e.g. the level or the temperature) may be better controlled than those of the river.

- Having multiple water intakes separated by sufficient distance, or with different physical position ensuring different access to the UHS[36].

(b) Air as UHS

Using the atmosphere as a heat sink offers several means, as the heat may be evacuated either through evaporation or in dry mode:

i. The principle of evaporation consumes water, which needs to be refilled when the autonomy is exceeded. Examples of use of this principle are given below:
    - Wet cooling tower: steam is produced at a lower temperature than as described in (a) above and the flow of water which is evaporated is slightly reduced. The water temperature is in a range comparable to what is obtained with the cooling provided by a river or a lake. The water needs to be circulated, and fans may also be necessary to improve the efficiency.
    - Production of steam at the surface of a pool, generally at atmospheric pressure: this is a typical backup strategy for SFP cooling, but it is also used for ensuring long term core cooling by placing a heat exchanger in a pool. The volume of the pool is designed to meet the targeted autonomy. A make-up is necessary at some point, when the initial amount of water is fully transformed in vapour, but this is typically a long term need.

ii. Using the atmosphere in dry mode needs greater exchange surface or flow than with evaporation, and provides cold water at a temperature which is significantly higher than a wet cooling tower. This has to be considered in the design of SSCs to be cooled, which should tolerate higher temperatures. The risk of clogging of air cooled systems is generally limited but cannot be excluded, for instance in case of a sandstorm, or in case of frost and/or icing, depending on the site, particularly for wet cooling towers, more exceptionally in case of volcano ashes reaching the site. Also, air cooled systems in dry mode may need to be protected against missile type hazards.

Whatever the cooling method, the loss of the heat sink may also arise from failure of the SSCs transferring heat towards the heat sink. Generally, all new NPP projects consider the loss of the main heat sink (where different from air) due to the above-mentioned hazards as a possible PIE and provide a combination of diverse UHS and diverse access to the UHS. The heat sink and the pathway to the heat sink may vary for the different plant states.

7.2.2. Different access to the ultimate heat sink

7.2.2.1. PWR and WWER

This paragraph will focus on typical applications for PWRs and WWERs.

Several options for access to the heat sink are typically used during DBA:

- If cooling by SGs is available, the cooling of the reactor is similar to the transient from power to shutdown mode in normal operation mode. Once appropriate conditions are reached, the RHR system is eventually connected to the primary circuit, allowing the core residual heat to be transferred by the cooling chain RHR system–component cooling water system (CCWS)–essential service water system (ESWS) to the UHS.

---

[36] For example, in France the EPR design features a backup water intake in the normal water outfall.

- If cooling from the secondary side is not available (e.g. use of feed and bleed strategy), the residual heat is removed from the core by the ECCS and from the containment by the CSS, if part of the design. Generally, at least one of these systems is cooled by the cooling chain CCWS–ESWS.

A typical case of a DEC without significant fuel degradation is the loss of the main UHS. In this case, several alternative options for access to the UHS may be used for heat removal from the reactor core:

- If the SGs and the emergency feedwater system are both available, the atmospheric steam dump system is used to transfer the core residual heat to the atmosphere as an alternative UHS as long as the RHR system is not available. Sufficient water volume is needed to reach the expected autonomy;
- If the SGs are available while the emergency feedwater system is not, the passive residual heat removal system from the secondary side is used to transfer the core residual heat to the atmosphere as an alternative UHS (this is the case for CNNC's HPR1000, which features both active and passive systems);
- If the SGs are not available, the residual heat may be transferred from the core to the reactor coolant and then to the containment using a feed and bleed strategy or a passive residual heat removal system for NPPs equipped with fully passive systems (e.g. the Westinghouse Advanced Passive 1000 MW, AP1000) and then removed from the containment using the containment heat removal system, which may be either active or passive.

In case of a severe accident, heat is initially transferred from the core to the containment atmosphere through a breach in the reactor cooling system (or the RPV itself in case of successful IVMR). Then, it is removed from the containment atmosphere using the containment heat removal system, which depending on the reactor design:

- It may be the same system for both categories of DECs, as for instance for the WWER-TOI in the Russian Federation;
- Or it may be different and diversified, for instance in the case of the diverse CSS implemented on the APR1000 in the Republic of Korea.

In the longer term, and in case of an ex-vessel reactor cooling strategy (for APR1000, EPR, WWER-1200), the containment heat removal system also evacuates heat from the core catcher.

7.2.2.2. ABWR

In the ABWR, the following systems are used to access the UHS:

a) RHR system, reactor building cooling water system (RCW), and reactor building service water system (RSW): these systems are used to access the UHS during normal operation and design basis fault conditions. The UHS ensures that an adequate source of cooling water is always available for reactor operation, shutdown cooling and accident mitigation. The RSW receives the cooling water from the UHS and returns water to it. The conceptual configuration of the SSCs related to the UHS can be summarized as follows;
   - UHS;
   - The heat exchanger building houses the RSW pumps, associated piping and valves;
   - The RSW is divided into three independent and separated divisions, each one provided with three RSW pumps, the associated piping and valves, instruments and

controllers;

- The RCW is divided into three independent and separated divisions, each one provided with three pumps, three heat exchangers, and the associated piping, valves, instruments and controllers.

b) The alternate heat exchange facility is used to recover the cooling capacity of any division of the RHR system by connecting NPE to the RCW in case of failure of the RCW or RSW system.

In case of loss of the normal cooling, the SFP may generally be cooled by an alternative system to avoid boiling. In both cases, the UHS may be the air or the water. In case of total loss of both systems, cooling may be ensured by boiling, while the SFP water level is maintained by a make-up system. To ensure the 'practical elimination' of fuel uncovering, this make-up system is generally diversified, and may use permanent and non-permanent sources of water.

7.2.3. Demonstration of protection against levels of natural hazards exceeding those considered for design

The topic was addressed in Section 6.2.2, which illustrates various complementary approaches used to demonstrate the presence of adequate margins to protect items ultimately necessary to prevent an early radioactive release or a large radioactive release in the event of levels of natural hazards exceeding those considered for design.

Natural hazards that are relevant for the UHS are site dependent and usually include earthquake, water level (minimum and maximum), water temperature, and air temperature.

8. SUPPLEMENTARY USE OF NON-PERMANENT EQUIPMENT FOR ACCIDENT MANAGEMENT

While the IAEA Safety Glossary [7] does not provide a definition for 'non-permanent equipment', SSR-2/1 (Rev. 1) [1] refers to 'non-permanent equipment' in footnote 22 in para. 6.28B which reads: "non-permanent equipment need not necessarily be stored on the site".

Furthermore, IAEA Safety Standards Series No. SSG-54, Accident Management Programmes for Nuclear Power Plants [59], states in the footnote in para. 2.21 that "'non-permanent equipment' is portable or mobile equipment that is not permanently connected to the plant and is stored in an on-site or an off-site location".

Therefore, NPE may indicate either:

a) Equipment on the site but not permanently connected:
   - Equipment permanently installed, ready to be connected;
   - Equipment stored on-site, that can be brought and connected;
b) Equipment off the site:
   - Equipment stored near the site but in a remote location;
   - Equipment stored in a centralized centre, requiring a certain time to be deployed on-site.

Depending on the country, but also on the location and accessibility of the NPP, either one or the other option, or a mix of these possibilities, may be favoured. Furthermore, the size of the equipment has an impact (for example, bringing a DG to supply a safety injection pump will likely require considerable efforts, especially if the normal access to the plant is no longer available), hence, different types of equipment may be defined accordingly: light or heavy equipment, portable (i.e. for a single human) or transportable (i.e. with a tool or vehicle).

8.1. RELEVANT IAEA SAFETY STANDARDS

The following subsection present an overview of related requirements in SSR-2/1 (Rev.1) [1] and recommendations from related safety guides.

8.1.1. Requirements from SSR-2/1 (Rev. 1)

Requirements related to the use of NPE for accident management are summarized hereafter:

- SSR-2/1 (Rev. 1) [1] Section 6 (design of specific plant systems): Requirement 58 (control of containment conditions), where a definition is provided for 'non-permanent equipment', Requirement 68 (design for withstanding the loss of off-site power), Requirement 80 (fuel handling and storage system).

8.1.2. Recommendations and guidance from SSG-2 (Rev. 1) on how to comply with the safety requirements

Paragraphs 7.51 and 7.64 of SSG-2 (Rev. 1) [13] provide recommendations pertaining to the crediting of NPE in the safety demonstration in both categories of DECs.

8.1.3. Recommendations and guidance from SSG-54 on how to comply with the safety requirements

SSG-54 [59] refers extensively to the use of NPE in the framework of accident management programmes, including in relation to: necessary steps to be adopted by operators for installation and operation, and availability of support items such as fuel (para. 2.18); connection to the plant to preserve the fundamental safety functions (para. 2.28); guidance and testing (paras 2.56, 2.59, 2.83, 3.2–3.3); location and hook-up points in relation to external hazards (paras 2.63–2.64, 2.82); use of NPE shared among more units (para. 2.74); training (para. 3.114).

8.2. EXPERIENCE IN THE IMPLEMENTATION OF SSR-2/1 (REV. 1) REQUIREMENTS

The following subsections illustrate practices in the consideration of NPE in the safety approach (subsection 8.2.1), typical examples of NPE (subsection 8.2.2), practices in ensuring access to connecting points (subsection 8.2.3) and in ensuring the required quality and availability of NPE (subsection 8.2.4).

8.2.1. Non-permanent equipment in the safety approach

SSR-2/1 (Rev. 1) [1] formulates requirements for NPPs to foresee "features to enable the safe use of non-permanent equipment"[1] for the following functions:

- Restoring the capability to remove heat from the containment (para. 6.28B of SSR-2/1 (Rev. 1) [1]), to avoid overpressure;
- Ensuring sufficient water inventory for the long term cooling of spent fuel and for providing shielding against radiation (para. 6.68 of SSR-2/1 (Rev. 1) [1]);
- Restoring the necessary electrical power supply (para. 6.45A of SSR-2/1 (Rev. 1) [1]).

However, SSR-2/1 (Rev. 1) [1] does not define further requirements for these "features to enable the safe use of non-permanent equipment", nor their role in safety analyses for DBAs or DECs.

Paragraphs 7.51 and 7.64 of SSG-2 (Rev. 1) [13] recommend that NPE is not considered in demonstrating the adequacy of the NPP design and that such equipment is typically considered only for long term sequences in accordance with the emergency operating procedures or accident management guidelines (as addressed in SSG-54 [59]). Paragraph 7.64 of SSG-2 (Rev. 1) [13] also states that "the time claimed for availability of non-permanent equipment should be justified".

In practice, as most of the systems ensuring the functions listed in SSR-2/1 (Rev. 1) [1] are closed cooling circuits or electrical systems, the features to enable the safe use of NPE generally include connecting points to some existing permanent SSCs, and procedures for their implementation: the consideration of relevant operational experience from the use of NPE is key to ensure that the interfaces between installed equipment and NPE are appropriately designed.

National regulations are very different about requirements on time availability of NPE, some being very precise on possible claims, other expressing high-level requirements while requesting a justification of any claim. In addition, in the NPP designs considered in this TECDOC, some countries do not make a distinction among light NPE (that may be easily implemented by a few individuals) and heavy NPE (i.e. China, Republic of Korea) while some other countries do (i.e. France[37], USA).

Unless when precisely defined by a regulation, the use of NPE generally follows a graded approach, specific to the site, the design, and the country:

- In the **short or very short term**, only equipment already installed (but not yet connected) or light NPE is allowed. Short term is country-dependent (typically less than 8 hours in the Republic of Korea — however 24 hours for the design of the APR1000 — and less than 6 hours in France). Light equipment is typically stored on a centralized location (serving multiple units on the same site) or on each unit. Light equipment may also be brought from off-site, provided that appropriate arrangements are in place or with an additional necessary time (typically less than 24 hours).
- In the **medium term**, the use of equipment of a larger size, located on-site or off-site, is allowed. Medium term is in the range of 24–72 hours. That equipment may require several individuals (depending on their skills and training) for its implementation. The size of such equipment is limited by the accessibility to the site and the transportation capabilities.
- In the **long term** (after 3 to 6 days), heavy equipment may be brought from off-site after full restoration of the access to the site or with additional preparation activities for the connection.

In this graded approach, additional rules and criteria may also be considered, for example the safe state should be reached using only permanent equipment, and NPE may be used to maintain it. More details on the specific approach taken in the design of the APR1000 are presented in Annex IV.

Alternatively, or in complement to this graded approach, some countries require that NPE capable to fulfil the specified functions should be connected to the plant in the short term, typically in less than 6 hours. This is the case, for example, of CNNC's HPR1000, for which the emergency water supply pump (which is connected to ad-hoc connecting points to the SIS to inject emergency water into the primary system) should meet the requirement of removing the reactor residual heat 6 hours after the shutdown. The same requirement exists for the secondary circuit, where the emergency water supply flow rate also meets the requirement of removing the core residual heat 6 hours after the reactor shutdown. Both these requirements have been introduced by the Chinese regulator after the Fukushima Daiichi accident and corresponding design features were implemented in CNNC's HPR1000 design, even though the use of NPE is not necessary for the first 72 hours after the onset of a severe accident.

In practice, provisions are in place to transport the NPE, such as dedicated trucks, which are stored on site or in its immediate vicinity. For example, the ABWR is equipped with NPE to restore the necessary electrical power supply that is installed on a large power truck, to recover the cooling capacity of any division of the RHR system by supplying power in the event of a 'LOOP + Class 1 EDGs failure'. The truck is equipped with multiple connection points.

Requirements may also be defined in national regulations regarding the need for NPE and related connecting points in terms of design, construction, maintenance, tests, drills, accessibility, etc. In the United States, for example, the U.S. NRC Regulatory Guide 1.226 [60]

---

[37] Through requirement on their deployment time.

endorses with clarifications NEI 12-06 [61] that outlines an approach for adding diverse and flexible mitigation strategies to address an extended loss of AC power (ELAP) and loss of normal access to the UHS occurring simultaneously at all units on a site.

The implementation of NPE, even if it is not required, may improve the accident management capability by the provision of diverse means. For instance, in Japan, the ABWR foresees NPE for restoring the capability to remove heat from the containment, which is an alternate heat exchange facility, while the permanent equipment is the filtered containment venting system, which copes with the loss of the RHR system.

In general terms, NPE is designed to provide additional flexibility to face events that exceed the design basis of the plant, as part of the lessons learned from the Fukushima Daiichi accident, where the three functions listed above[38] may have been lost. The definition, design and implementation of features to enable the safe use of NPE is generally made such as to be designed (or protected, considering their location):

- Against design basis hazards;
- As well as against levels of natural hazards exceeding those considered for design, for some specific beyond design basis natural hazards (see Section 6 for more information on this topic).

Hence, external hazards affecting the plant will not fail the NPE wherever it may be located.

Such features have also to be assessed for their potential detrimental effects (due to spurious or erroneous (operator error) actuation) to effectively make a balanced decision via the consideration of the different reasonably practicable options; this may include the provision of permanent equipment.

NPE are designed to be able to be connected to a specific part of the plant and can only be used in the range of its technical and operational limits, as defined in its technical specification or supplier notice for use. An example to illustrate such limitations is the design of a pump to feed into SGs. Adding a connecting point to the SGs feedline is a relatively simple task, provided that appropriate isolation devices are in place and that this is not leading to an excessive increased risk of failure of normal and emergency feedwater systems or adding new initiating events. The events in which the non-permanent pump would be used need to have limited effects on the SG's capability. The SG's pressure in these conditions needs to be clearly defined for inclusion in the pump specification. A low-pressure pump would be easier to design/provide but requires controlling the SG's pressure. This consideration may be a first driver to decide whether adding a connection point is worthwhile and practicable.

8.2.2. Examples of non-permanent equipment

The necessary connecting points allowing fulfilment of the three functions required by SSR-2/1 (Rev. 1) [1] are generally available in all NPP designs considered in this TECDOC. They are generally supplemented by, at least, one function aimed at removing heat from the reactor cooling systems (hence aiming to preventing core melt).

In practice there is not a unique list of non-permanent means shared among the various NPP designs, as these are highly event-dependent, reactor technology dependent (i.e. the cooling

---

[38] i) removing heat from the containment; ii) restoring an electrical power supply; iii) restoring cooling in the SFP.

requirements are clearly quite different for a large size reactor and a very small size reactor) and site-specific.

In some cases, the restoration of these functions may be achieved by the provision of light equipment or a small DG. Whatever the requirements, non-permanent electrical supplies may use different types of generators, in terms of power and voltage. Examples depend on the design and the function to be restored:

- Limited power generators (less than 1.0 MW in the Republic of Korea) may be used to supply electrical power for battery chargers, motor-operated isolation valves, mobile air compressors, temporary fans, ignitors, main control lighting, instrumentation, etc.
- Higher power generators[39] (up to 3.2 MW in the Republic of Korea) are needed to supply cooling pumps (such as auxiliary feedwater pump, emergency core cooling pump or residual heat removal pump), motor-operated valves, heating, ventilation and air conditioning (HVAC), etc.

8.2.3. Ensuring access to connecting points

The connection of NPE to connecting points faces several challenges:

- NPE located on-site needs to be protected against hazards, including those which might have caused the loss or failure of permanent equipment;
- Depending on the hazards (i.e. extreme natural hazard), damages may have occurred to the site access (bridges or roads may be blocked or damaged), or the site may be isolated (i.e. flooding), challenging the site access for off-site equipment;
- Similarly, extended damages may occur on the site itself, and (depending on the progression of the accident) the radiological risks may limit any human intervention;
- Finally, in extreme situations the connecting points may be affected, in case these were protected against a lower level of hazard than the one having caused the accident conditions.

Several practices are adopted to face these challenges:

a) When NPE is stored on site, then an appropriate building, which may be specific to each unit or shared for all the site units, is used:

  - In China and the Republic of Korea, for example, such a building is located at least 100 m away from safety related buildings and designed against site specific external hazards such as earthquake or flooding. For instance, in a seaside location, it may be located 5 m higher than the design basis flood level and designed with an additional margin regarding the seismic basic intensity in the site area in China.
  - In the case of the Fuqing NPP site, which contains six units (two of which are CNNC's HPR1000 reactors), two mobile emergency water supply pumps are shared by all the six units: therefore, in the case of a simultaneous accident to all six units, a priority between the units is necessarily be defined for the use of mobile emergency water supply pumps.

---

[39] In the Republic of Korea, the 3.2 MW mobile generators are commercial grade and non-seismic, but the pre-staged foundation, fuel tank, and dike are seismically qualified.

b) In case of external hazards, even if site access roads having previously been affected by an extreme natural event can be repaired by heavy machinery stored in various and diversified locations, this may take some time, and adequate transportation means needs to be provided for bringing equipment from off-site. These could include reinforced trucks, sea and river barges, and helicopters.

c) To face damages on the site, heavy means to remove debris may be provided, and several connecting points foreseen, and physically separated. More generally, connecting points are protected against external hazards and generally requirements for their seismic design are the same as for the systems to which they are connected. Special protective equipment or apparatus is provided to protect workers against exposure to ionizing radiation, and the design of connecting points and the implementation procedures minimizes the necessary connection time.

8.2.4. Ensuring quality and availability of non-permanent equipment

As NPE is not credited in the safety demonstration (except for long term sequences and in accordance with the emergency operating procedures or accident management guidelines, see paras 7.51 and 7.64 of SSG-2 (Rev. 1) [13]), it is generally designed as non-safety classified and may be categorized as non-seismic. On the one hand, this will allow for design and manufacture following non-nuclear industrial standards, bringing more flexibility for the supply chain and even for bringing some to the accident site during an emergency, as any similar equipment may fit. On the other hand, however, this may limit its use or require storing it in a remote location, to protect it from the hazards affecting the site.

Consequently, NPE and corresponding connecting points are standardized as far as possible for a given site (even in the case of different designs on the same site) and at a national level. When standardization is not possible, additional equipment is provided, however, the possibilities to share it between different sites or units are reduced.

In the extremely unlikely case that more than one heavy equipment of the same type is necessary, it has to be brought from another site, and considered as off-site equipment. The prioritization of the use of on-site NPE is made by the emergency team, considering at least the deterioration status of the units and the necessary time for NPE to be brought on-site.

NPE, connecting points and associated isolation devices and lines are maintained, periodically inspected and tested. Training programmes are in place to ensure the efficiency of the implementation of NPE and associated procedures. For example, in the Republic of Korea, the nuclear regulatory body requires the first training to be performed within the first year after the approval of the accident management plan. Periodic training has eventually to be performed every two years. In case of major modifications to the coping strategy, the licensee needs to set up a training plan within three weeks and complete the training within six months. Each training may be accompanied by a radiological emergency preparedness drill.

Appendix

EXAMPLES OF DEMONSTRATION OF 'PRACTICAL ELIMINATION' OF SEQUENCES THAT COULD LEAD TO EARLY OR LARGE RADIOACTIVE RELEASES

A.1. CNNC'S HPR1000: 'PRACTICAL ELIMINATION' OF EARLY OR LARGE RELEASES (CHINA)

A set of reliable and effective mitigating measures have been designed for CNNC's HPR1000 to prevent and mitigate severe accident phenomena (e.g. HPME, DCH, hydrogen detonation, steam explosion, containment overpressure, basement melt-through) that may lead to large releases. These are the dedicated depressurization system, containment hydrogen combination system, cavity injection and cooling system, and the PCS.

The availability of equipment and instruments required for severe accident management are evaluated by comparing the severe accident environmental conditions with the qualification conditions. Typical accident sequences are selected to evaluate the effectiveness of the mitigation measures. The frequency in the 95th percentile of the conditions that lead to large releases is lower than 10^{-7}/r.y.

The potential radiological consequences of the plant event sequences with a significant frequency of occurrence are demonstrated to meet the safety objective in the case of a severe accident, which is that only protective actions that are limited in terms of lengths of time and areas of application would be necessary and that off-site contamination would be avoided or minimized.

The CDF of CNNC's HPR1000 in the shutdown state with open containment is very low. When an accident occurs in the shutdown state with open containment, the operators can close the containment penetrations (valves and hatches) rapidly according to the operating procedures to prevent early or large releases.

There are no safety systems and safety features designed to mitigate the failure of a large component in the RCS. Several sets of well-established technical standards are available to ensure high quality and reliability of large pressure vessels. The 'practical elimination' of failures of large components is thus achieved by the essential means of the DiD level 1 without relying on the subsequent levels of DiD.

For uncontrolled reactivity accidents of CNNC's HPR1000, the main protection is provided by ensuring a negative reactivity coefficient with all possible combinations of the reactor power and coolant pressure and temperature.

Measures, which are feed-bleeding of the RCS, heat removal of secondary side, dedicate depressurization, are designed to depressurize the RCS and prevent SGTR. The frequency of occurrence of such an accident is very low. The early large release frequency in the 95th percentile associated to ISLOCA event is lower than 10^{-7}/r.y.

A.2. EPR: 'PRACTICAL ELIMINATION' OF HIGH-PRESSURE MELT EJECTION (FRANCE)

A.2.1. Requirements considered for the reactor design

The implemented solutions were designed to fulfil the approach recommended by the Technical Guidelines [62] and reproduced hereafter:

"[…] A design objective is to transfer high-pressure core melt sequences to low pressure sequences with a high reliability […].

This objective implies to limit the pressure in the RCS in the range of 15 to 20 bar, when the reactor pressure vessel rupture may arise. This objective can be ensured by adding, to the depressurization function of the pressurizer valves, a dedicated bleed valve with an isolation valve […].

The discharge capacity of the dedicated valve must be determined considering the following situations, with realistic assumptions:

- Loss of off-site power with unavailability of all DGs;
- Loss of off-site power with unavailability of all DGs but with recovery of water supply during core melting;
- Total loss of feedwater combined with the failure of the primary feed and bleed[40].

However, sensitivity studies regarding the discharge capacity, the hot gas temperatures and the opening criteria must be performed by the designer considering delayed bleeding and late reflooding as well as the uncertainties of the code models related to the late core degradation phase or reflooding. These sensitivity studies will also assist in determining the way of actuation of the dedicated valve (manual or automatic), considering the possibility of human errors during the accident.

The dedicated valve and its isolation valve must be qualified under representative conditions. Experimental justifications may be necessary, especially for those conditions that deviate considerably from normal operating conditions.

On another hand, design provisions must be taken to cope with the mechanical loads which would result from the reactor pressure vessel failure at 20 bars to limit the vertical upward movement of the reactor pressure vessel.

Moreover, design measures must be taken to limit the dispersal of corium into the containment atmosphere in the event of a reactor pressure vessel melt through, to prevent direct containment heating. These design measures are related to the reactor pit and its ventilation as well as to the ex-core neutron

---

[40] It is supposed that the pressurizer valves are not available; the dedicated valve and its isolation valve remain available.

measurements, to ensure that large quantities of corium released from the reactor pressure vessel cannot be carried out of the reactor pit".

A.2.2. Implemented provisions and related supporting studies

A.1.2.1. Implemented solutions

To achieve the above objective (i.e. to limit the pressure of the primary circuit to a sufficiently low value at the time of a potential vessel rupture), the RCS is equipped with an ultimate discharge system (dedicated bleed valve with an isolation valve), different from the overpressure protection of the pressurizer (safety relief valves, SRVs). This discharge system consists of two parallel discharge lines. Each line has two valves in series (one of them is an isolation valve) qualified to severe accident conditions.

Both lines have the same discharge capability and fulfil the following two functions:

- Feed and bleed function (feeding is provided by the injection system) to prevent core melt;
- A fast discharge of reactor coolant in the early phase of a severe accident.

Two batteries from two separated divisions, implemented for severe accident conditions, each supplying a series of two valves, ensure the actuation even in case of a loss of off-site power with a combined failure of the six backup DGs (four EDGs and their two ultimate DGs, also called 'SBO DGs').

Each of the two lines has a discharge capability allowing to limit the pressure of the RCS to an appropriate low pressure (severe accident low pressure criterion) at the time of the pressure vessel rupture. The line is manually opened by the operator based on the severe accident criterion of core outlet temperature.

Regarding the potential effects of a DCH, they are mainly driven by two initial conditions influencing the core melt, its fragmentation and its dispersal from the reactor pit to the upper part of the containment building:

- Thermal-hydraulic conditions, in particular the pressure within the RCS at the time of the pressure vessel rupture;
- The design (geometry) of the reactor pit and the openings towards the containment dome.

The reactor pit geometry is designed with only small openings to the primary loop compartments, without direct path to the upper volume of the containment. This design favours corium retention in the lower zone of the containment and reduces the volume of gas involved in the thermal exchanges. The low pressure criterion for severe accidents is a decoupled value to justify the avoidance of any damage to the containment resulting from the dispersal of the corium.

A.1.2.2. Probabilistic assessment

A probabilistic analysis is performed to show that the frequency of occurrence of a sequence of HPME is compliant with the probabilistic target set by the designer to consider a condition as 'practically eliminated'.

In addition to the contribution of the provisions described in the previous sections of this Appendix, the level 2 PSA supporting studies show that a hot leg rupture is very likely to happen before the pressure vessel failure, thus contributing to the reactor coolant depressurization.

A.3. ABWR: 'PRACTICAL ELIMINATION' OF FUEL DAMAGE IN THE SPENT FUEL POOL (JAPAN)

The demonstration of 'practical elimination' for ABWR plants is principally based on both the consideration of the diverse design measures and DiD provisions in the design and supported by results from extensive PSA and severe accident analysis. The demonstration is applied to three fault groups: reactor power operation, shutdown reactor condition, and SFP faults. An illustration of a three-step approach in the demonstration of 'practical elimination' of early or large releases from severe accidents in the SFP is summarized in Sections A.3.1 to A.3.3 below.

A.3.1. Step 1: Identification of design provisions

The following are the design measures and features available in the ABWR for the prevention and/or mitigation of a severe accident in the SFP:

1. Large volume of water in the SFP: as a large amount of water pool is present in the SFP, there is a large time margin to prevent fuel damage in the SFP. The time margin to reach the top of the active fuel is estimated to be more than 300 hours. Even in case of a small LOCA, the time margin to reach the top of the active fuel is estimated to be about 50 hours.
2. Fuel pool cooling and clean-up system: the fuel pool cooling cools the SFP by removing the decay heat from the spent fuel and maintains the temperature below the design values.
3. RHR: the RHR provides the fuel pool cooling with supplementary cooling to maintain the SFP water temperature within the design values by removing decay heat in the event of full core unloading where the heat load to the pool exceeds the fuel pool cooling capacity.
4. Make-up water condensate system: the make-up water condensate system supplies water to the SFP from the condensate storage tank to compensate evaporation during normal operation. It is also designed to have sufficient make-up volume to compensate leakage from liner cracks or overflow due to any event such as earthquake.
5. Suppression pool clean-up system: the purpose of the suppression pool clean-up system is to clean up the water in the suppression pool by transferring the pool water through the filter demineralizer of the fuel pool cooling and clean-up and returning it back to the suppression pool.
6. Fire protection system: this system is used as an accident management measure in case of multiple failures of other mitigating features. Among three fire protection pumps, one diesel driven pump is available without AC and DC power.
7. Flooder system of specific safety facility (FLSS): the FLSS is an additional safety feature for core damage prevention and mitigation. It is installed in the backup building (also called 'specialized safety facility' in Japan) and can supply water to the SFP from a water source. The FLSS fulfils not only the function of water filling but also the function of water spray. The spray header will be installed on the peripheral edge of the pool above the normal water level of the SFP. If water level cannot be recovered, the water spray will spread over the whole surface of the SFP, allowing cooling of fuel bundles in the SFP by the water spray.
8. Flooder system of reactor building (FLSR): the FLSR can supply water to the SFP from a water source by mobile pumps. The FLSR fulfils not only the function of water filling but also the function of water spray.

Based on the above diverse design features and measures that are available, the ABWR design can 'practically eliminate' the risk of early or large release initiated by faults on the SFP. The justification is provided through PSA results, further described in Step 3.

A.3.2. Step 2: Identification of representative severe accident scenarios

The 'conditions' leading to an early release and a large release were derived on the basis of a review of the release categories, the representative severe accident scenarios, and the results from the level 2 PSA.

In the level 2 PSA, several risk metrics were evaluated and these included containment failure frequency, large early release frequency and LRF. The results of the level 2 PSA analyses, along with those of the level 3 PSA, were compared against the numerical targets identified in Step 3.

The contribution from events initiated by external hazards was excluded from this demonstration, based on the recommendation from IAEA-TECDOC-1791 [9]:

- For some external hazards, it may not be practical or even possible to demonstrate that the occurrence of a hazard of such severity that could cause extensive plant damage leading to a large or early radioactive release, and therefore needing to be 'practically eliminated', is below a threshold such as 10^{-6}/r.y.
- This shows the limitations of probabilistic methods to claim the demonstration of the 'practical elimination'. For this reason, it is advisable to keep the 'practical elimination' concept for external hazards separate from those associated with internal plant sequences.

The following are representative severe accident scenarios for the SFP:

a) Water boil-off: in this scenario, all heat removal systems and all water injection systems for the SFP are assumed to fail. Hence, water evaporates gradually due to the decay heat from the spent fuel and, consequently, the water level of the SFP gradually decreases. As a large amount of water is present in the SFP, there is a long-time margin prior to fuel damage.

b) Small LOCA: in this scenario, loss of the SFP water inventory is assumed due to the SFP liner failure. As all leak flow goes through the leak detection lines, the operator can respond to the alarm and isolate the line. However, the isolation of the leak detection lines is assumed to fail. In addition, all water injection systems for the SFP are assumed to fail. Hence, the water level decreases below the top of active fuel and fuel damage occurs.

c) Catastrophic failure: in this scenario, the SFP water level is assumed to decreases rapidly due to the large loss of the SFP water inventory (large leak). The leak rate is not specified but it is assumed to be larger than the design leak rate and larger than the water injection flow rate by FLSS and FLSR. Therefore, the water level cannot be recovered. This scenario is assumed to occur due to cask drop on the SFP edge and is considered in the internal hazards PSA for the SFP.

A.3.3. Step 3: Demonstration of 'extremely low likelihood' with a high degree of confidence

The internal event PSA results for the representative severe accident scenarios in the SFP as well as the associated time margins for a large release are provided in TABLE 11.

The PSA results were then compared against the following numerical targets:

- LRF of 10^{-6}/r.y;
- A value of 10^{-7}/r.y for individual release categories;
- Other numerical targets required from regulations in each country (e.g. basic safety objective targets 8 and 9 in case of the United Kingdom).

All four numerical targets were achieved for the SFP.

TABLE 11. PSA RESULTS FOR LRF FOR INTERNAL EVENTS IN THE SFP

Facility	Accident scenarios	Internal event PSA result	Time margin for a large release
SFP	Boil-off / LOCA (Late release)	4.8E-08/r.y	300/50 hours
	Catastrophic failure (Early release)	8.8E-11/r.y	> 0 hours

The scope of PSA performed for ABWR also includes internal fire PSA and internal flooding PSA. The current integrated PSA results show that the risk of fuel damage from fire, flood and seismic hazards dominates the risk results. The preliminary nature of ABWR generic design without considering detailed design information (e.g. layout of cabling) has resulted in the hazards assessments being based on more conservative assumptions than those in the internal events assessment.

For the demonstration of 'practical elimination' of sequences potentially leading to large or early radioactive releases, initiated by internal events associated with internal hazards, the following points are relevant:

- No release categories leading to severe accidents initiated by hazards were identified through PSA, and the release categories for the hazards PSA are adequately represented by the equivalent release categories defined for internal events.
- For this demonstration, the design provisions summarized in Section A.3.1 of this Appendix are similarly applicable for the prevention and mitigation of severe accidents initiated by hazards.

Based on the results, and again assuming a numerical threshold of 10^{-6}/r.y for the LRF for the demonstration of 'practical elimination', the following conclusions can be drawn:

- The LRF calculated from the internal hazard PSAs (fire and flood) is below the threshold with a certain margin.
- The aggregated LRF for severe accidents initiated by internal events, including internal hazards, is also below the threshold.

Hence, these results provided further evidence in the demonstration of 'extremely low likelihood with a high degree of confidence'.

A.3.4. Discussion

In both scenarios of water boil-off and small LOCA, there is a large time margin available to undertake mitigation. It takes more than 300 hours and approximately 50 hours for the water level to reach the top of active fuel in the boil-off scenario and small LOCA scenarios, respectively. In the catastrophic failure scenario, the time margin is much smaller.

However, the probability of occurrence which may lead to early release is negligibly small in comparison with numerical targets. In addition, the PSA was based on the following conservative assumptions:

- The decay heat of the fuel bundles is conservatively assumed to be equal to four days, which corresponds to the maximum decay heat in the SFP. It will lead to an earlier fuel damage and associated early release.
- The recovery of the RHR by a large power truck which is available in the ABWR design was not considered. Sensitivity analysis based on RHR recovery showed a reduction in the LRF of 43%.
- The recovery of the FLSS was not considered. The FLSS may be fixed within the long timescales that are available.
- The success probability of implementing the FLSR is only 0.3 due to human error despite the long timescales that are available.
- There is no off-site support considered (e.g. fire trucks).

Therefore, if recovery of safety systems and off-site support are considered, the CDF of the boil-off and small LOCA scenarios would be considerably smaller than as obtained through current PSA results.

It can be concluded that ABWR 'practically eliminates' severe accident scenarios that could occur in the SFP which could then lead to early or large releases, based on the diverse design features as well as demonstrating a very low likelihood occurrence through PSA results.

A.4. APR1000 (REPUBLIC OF KOREA)

A.4.1. Hydrogen explosion

For the 'practical elimination' of hydrogen detonation, which can threaten the containment integrity, severe accident dedicated PARs, also called HMS, are installed. Dozens of different sizes of PARs fulfil the function to prevent hydrogen detonation in the absence of power supply. The primary containment is designed to provide a large free volume and hydrogen flow paths so to facilitate hydrogen mixing and dispersion. The core catcher installed in the containment cavity also reduces the generation of ex-vessel hydrogen. The demonstration of 'practical elimination' of hydrogen detonation is achieved by demonstrating that this sequence is extremely unlikely, through conservative deterministic analysis considering uncertainty of the event sequences and the extremely low probability of containment failure due to hydrogen detonation, obtained through level 2 PSA.

A.4.2. Large steam explosion

The phenomenon of large steam explosion could occur due to FCI either inside or outside of the reactor vessel during core melt sequences, thus potentially leading to containment failure:

- In-vessel steam explosion: since the likelihood of early containment failure due to in-vessel steam explosion is known to be extremely low, it is considered as 'practically eliminated' by physical impossibility in the design of the APR1000. Despite the extremely low likelihood, deterministic in-vessel steam explosion analysis is performed to demonstrate the integrity of the RPV.
- Ex-vessel steam explosion: the ex-vessel FCI induced large steam explosion is prevented by the absence of water in the cavity at the time of vessel failure, in accordance with the post-flooding strategy of PECS, which contains the core catcher. Even though the ex-vessel FCI is excluded by keeping a dry cavity, the very unlikely scenario of pre-flooding of core catcher is assumed in the conservative deterministic analysis of energetic FCI and the analysis demonstrates that the integrity of reactor cavity structures is preserved. Supplementary PSA for the demonstration of 'practical elimination' of a large steam explosion is not used due to large intrinsic phenomenological uncertainties in the steam explosion.

A.4.3. Direct containment heating

DCH due to HPME in the APR1000 is prevented by the installation of the ERDS. This system consists of severe accident dedicated valves which are opened manually to reduce the RCS pressure below 20 bar before the RPV fails after a core melt sequence. The 'practical elimination' of DCH in the APR1000 is demonstrated by both deterministic and probabilistic approaches:

- A conservative deterministic case study is performed to demonstrate the effectiveness of the ERDS. To further ensure 'practical elimination' of containment failure due to DCH, the deterministic analysis includes the case of ERDS failure.
- The reliability of the ERDS including the operator action is evaluated by level 2 PSA to demonstrate that containment failure due to DCH is extremely unlikely.

A.4.4. Large reactivity insertion

Uncontrolled large reactivity insertion may be caused by either sudden insertion of a cold or unborated water plug into the reactor core or due to reaching criticality during a severe accident sequence:

- In the APR1000 design, cases where a heterogeneous boron dilution would occur are examined and analysed to ensure that they would not cause any large reactivity insertion;
- For severe accident conditions, criticality calculations for the molten corium both inside and outside of the RPV are performed to demonstrate that the reactivity remains subcritical during the evolution of the of severe accidents. The 'practical elimination' of large reactivity insertion is demonstrated by the above deterministic evaluations.

A.4.5. Rupture of major pressure components

A sudden mechanical failure of a large component in the RCS, such as the RPV, SG or pressurizer, would cause a loss of core cooling, and potentially also damaging the containment.

The 'practical elimination' of the RPV failure is demonstrated by application of robust design code of American Society of Mechanical Engineers [63], appropriate in-service inspection and periodic safety review. A supplementary probabilistic fracture mechanic analysis is performed to ensure that the failure of RPV is 'practically eliminated'.

A.4.6. Containment overpressurization

The loss of containment heat removal followed by a failure of the primary containment after a severe accident is 'practically eliminated' in the design of the APR1000:

- The diverse CSS, which is a system independent from the CSS used for DBA conditions, is designed to have sufficient capacity to remove decay heat of the molten core and to remove fission products after a severe accident supported by the diverse power and cooling systems. These diverse systems are designed to withstand the RSEH and SBO conditions.
- The conservative deterministic severe accident analysis proves that the containment integrity is maintained, and the level 2 PSA provides the supplementary evidence of 'practical elimination' of large or early releases due to containment overpressurization.

A.4.7. Basemat melt-through

If the ex-vessel molten core is not adequately cooled and interacts with the concrete structure, it is possible to cause a basemat melt-through or a loss of containment integrity due to the generation of a large amount of hydrogen and other non-condensable gases. The core catcher system of the APR1000 prevents the occurrence of basemat melt-through. The core catcher has a sufficient area to collect the molten core materials from the breached RPV. The core catcher also provides cooling capability to quench molten core in a passive way. The conservative deterministic severe accident analysis for the APR1000 demonstrates the performance of the core catcher. The level 2 PSA provides additional supplementary evidence that the likelihood of basemat melt-through for the APR1000 is extremely low with a high level of confidence so that the situation is 'practically eliminated'.

A.4.8. Containment bypass

Based on the level 2 PSA, three types of containment bypass sequences that can lead to a large or early radioactive release are considered in the design of the APR1000:

- SGTR;
- Failure of the CIS;
- Interfacing system loss of coolant accident (ISLOCA).

The SGTR induced large release, which is one of the high large release contributors in the level 2 PSA, is 'practically eliminated' by the robust SG isolation design provisions and the extremely low probability of its failure. The main steam atmospheric dump valve block valves are designed to manually isolate the SG in cases of the failed or stuck open main steam atmospheric dump valves, which is one of the potential pathways of radioactive releases for SGTR. The main steam safety valves which represent another release path for SGTR, are designed and manufactured in accordance with ASME B&PV Code Section III [63] components such that the possibility of failure to close is minimized. This reliable design feature enables that the SFC for the fail-open main steam safety valves is not considered in the deterministic safety analysis. The main steam isolation valves, which represent another release path, are designed to be redundant to provide reliable isolation of the SG. Eventually, the results of the level 2 PSA for the group of SGTR induced large release sequences show that its frequency is extremely low enough to ensure 'practical elimination'.

The second type of the containment bypass sequences, i.e. failure of CIS, is 'practically eliminated' primarily by the robust design of containment isolation systems such as; redundant and automatic design of the isolation valves located both inside and outside the primary containment, fail-safe design, independent and diverse emergency power sources to close the valve against both LOOP and SBO, allocation of high safety classification, provision of the limiting conditions of operation in the technical specifications, and the periodic leakage test requirements. The level 2 PSA supports that the large release frequency for the group of CIS failure sequences is extremely low with a high level of confidence.

A similar approach is applied to the demonstration of 'practical elimination' of ISLOCA induced large releases. Systems that are connected to the RCS are identified and designed to withstand the high pressure such that radioactive releases through the connected systems are avoided. High-pressure alarms are installed downstream of the isolation valves to initiate a manual isolation. The frequency of the group of ISLOCA induced large release sequences for the APR1000 is sufficiently low to demonstrate the 'practical elimination'.

A.4.9. Severe accident during shutdown operation with open containment

During the Mode 5 operation of the APR1000 with the RCS closed, the containment equipment hatch may be open to permit transfer of equipment in and out of the primary containment structure during plant outages. If an accident occurs, credit is given to closing the containment to allow mitigation of the accident progression. From the reference plant, the containment equipment hatch should be closed in less than one hour as long as AC power is available. If AC power is not available (i.e. SBO), a small standby power source such as generator or battery is provided. This could be immediately used to power the winches that normally raise and lower the hatch. According to the reference plant, time to boiling of RCS inventory is more than at least 2 hours. Therefore, the containment equipment hatch can be closed before reaching acceptance criteria such as temperature and radiation levels inside the containment. For Mode 5, when the RCS is not closed, such as when the pressurizer manway is open, the containment equipment hatch should be secured in plant outages with at least four bolts per operating procedure. Therefore, the containment isolations, including equipment hatch, withstand the containment pressure during accidents.

A.4.10. Failure of the spent fuel storage pool

A large release of radioactive materials might occur if the cooling capability or structural integrity of the SFP is lost. The design provisions of the APR1000 to prevent such an event are as follows:

- The SFP structure is designed to withstand the consequences of internal and external hazards (including a commercial scale airplane crash).
- Siphoning of the pool water is avoided by locating the suction lines above the minimum required pool water level.
- A complete loss of SFP cooling is prevented by reliable redundant cooling and monitoring systems and by providing an independent diverse cooling train which is powered and cooled by diverse and independent systems from the systems used for normal operation. Additional external non-permanent water make-up is supplied in the event of extended loss of AC power.
- Mechanical failure of the spent fuels by dropping of a heavy object is prevented by the fail-safe design of the fuel handling machine and the overhead crane installed in the fuel handling area of the fuel building.

An independent PSA for the SFP is performed to ensure that the failure of the spent fuel in the SFP is extremely unlikely with a high level of confidence for the demonstration of 'practical elimination'.

A.5. WWER AES-2006: 'PRACTICAL ELIMINATION' OF HIGH-PRESSURE MELT EJECTION (RUSSIAN FEDERATION)

The demonstration of 'practical elimination' for new WWER plants is based on the deterministic analysis of the bounding severe high pressure scenario supported by probabilistic assessment of high pressure CDF. This scenario is assumed to cover all sequences leading to high pressure in the reactor at the beginning of core degradation.

A.5.1. Bounding high pressure scenario

The scenario considered here assumes rapid emptying of all SGs, which results in dependent failure of both the active emergency heat removal system and the diverse passive heat removal system (PHRS).

This scenario may occur due to the consequences of a seismic event with a magnitude significantly exceeding the one considered for the safe shutdown earthquake for which the non-occurrence of the following dependent failures at the NPP site cannot be successfully demonstrated:

- Damage of the outdoor switchgear, long term LOOP;
- Turbine hall failure (loss of the normal heat removal, destruction of the turbine building with fall of the roof onto the steam pipelines);
- Failures occurring on normal operation equipment and safety systems, independent from the hazard;
- Eventual small LOCA;
- Subsequent secondary events (e.g. a fire in the room of the 10 kV sections due to short circuits on the sections and in the cables);
- Stack-open failures of steam dump to the atmosphere, depending on the power supply.

Hence, on this basis, in order to obtain a bounding envelope high pressure severe accident scenario with a presumed early melting of the reactor core, a long term blackout with a leakage of four steam lines, which leads to a complete loss of heat removal through the secondary circuit (dependent failure of PHRS) can be considered.

A.5.2. Scenario modelling (1st stage)

The rapid SG dry out leads to the inefficiency of heat removal systems, causing a pressure increase in the RCS, leading to primary coolant discharge via the pressurizer SRV, progressively leading to uncovering of the core. Partial fuel melting occurs 3 hours after the onset of the scenario.

According to the severe accident management procedure, the operator opens all pressurizer pilot operated SRVs.

In case the pilot operated SRVs failed to open, the flow of hot gases into the RCS would cause heating of the metal of the connecting surge line, hot loops and SG tubes, with consequential decrease in strength and subsequent high-temperature creep.

Thermohydraulic calculations show, however, that SG tube heating is controlled by the cold loop seals that prevent start of hot gases natural circulations over the SG tubes: hence, this phenomenon might occur only if the seals were emptied during the first stage of the accident. Even though these phenomena have not been experienced, these were conservatively postulated in order to evaluate the fragility margins of abovementioned SSCs.

A.5.3. Model for stress analysis (steam generator pipe versus surge line)

The finite element model was applied to define the critical areas to perform the fragility analysis.

Hence, the curves that represent the maximum allowable stress were obtained.

Finally, the time of rupture was determined combining the loads obtained through the model and the maximum allowable stress.

A.5.4. Scenario modelling (2nd stage)

After approximately 3.5 hours, conditions for the rupture of the connecting surge line by the criterion of static strength are met, eventually similar conditions are encountered in the SG tube bundle if the connecting pipe does not break.

If the loop seals are not emptied and due to the consideration of a margin (in the limiting stresses parameter), the rupture of surge line will be prevented, conditions for the creep rupture of the hot leg are met, since the high temperature of the hot leg metal (which is the highest in the RCS) reaches a critical value based on the creep criterion.

In all cases, the depressurization of the primary circuit cannot be avoided, with consequential pressure reduction and further injection of borated water into the reactor, first from medium-pressure hydroaccumulators, then from low-pressure ones.

The final phase of fuel melting and the failure of the reactor vessel occur at low pressure, while the corium catcher is effective to provide corium cooling.

The containment failure (due to the pressure increase up to 0.7 MPa, which represents the limit in elastic mode) occurs about 36 hours later and the total containment failure due to overpressure occurs at longer time.

The result of consideration of bounding scenario demonstrates that high pressure sequences resulting in large early release can be considered as 'practically eliminated' in new WWER designs.

A.5.5. Robustness of the considered scenario

The demonstration of the robustness of the demonstration is important to guarantee appropriate confidence on the conclusions gained from the analysis of the abovementioned scenario.

For such purpose, the sensitivity of the scenario to possible alternatives at different stages of the accident was examined:

- Total loss of heat removal with filled SGs against emptying SGs: in this case, it was assumed the failure of both diverse systems, i.e. active emergency heat removal system and PHRS. Despite very low probability of such event, this assumption finally results in the same accident development (e.g. SG emptying due to steam dump to the atmosphere or SGs safety valves stacked open or steam line rupture if they fail to open). Should a cycling operation of safety valves take place, reactor cooling would be provided for longer time and consequently, the RPV will not fail at high pressure.

- Successful primary depressurization against operator failure: the second case is examined in the bounding analysis, while the first allows to maintain RCS integrity. Both lead to similar results from the standpoint of further severe accident analysis.
- Uncertainties and assumptions related to thermohydraulic calculations: the most important is the loop seal emptying phenomenon. The basic study uses conservative assumption.
- Surge line against SG tube rupture: these alternatives were analysed by finite element models.

A.5.6. Probabilistic assessment

The results gained from deterministic study are supported by probabilistic calculations performed in level 1 and level 2 PSA. In fact, all high pressure sequences, potentially leading to large early release, including those induced by external or internal hazards, have a total frequency lower than 10^{-7}/r.y.

REFERENCES

[1]   INTERNATIONAL ATOMIC ENERGY AGENCY, Safety of Nuclear Power Plants: Design, IAEA Safety Standards Series No. SSR-2/1 (Rev. 1), IAEA, Vienna (2016).

[2]   INTERNATIONAL ATOMIC ENERGY AGENCY, Governmental, Legal and Regulatory Framework for Safety, IAEA Safety Standards Series No. GSR Part 1 (Rev. 1), IAEA, Vienna (2016).

[3]   INTERNATIONAL ATOMIC ENERGY AGENCY, Safety Assessment for Facilities and Activities, IAEA Safety Standards Series No. GSR Part 4 (Rev. 1), IAEA, Vienna (2016).

[4]   INTERNATIONAL ATOMIC ENERGY AGENCY, Safety of Nuclear Power Plants: Commissioning and Operation, IAEA Safety Standards Series No. SSR-2/2 (Rev. 1), IAEA, Vienna (2016).

[5]   INTERNATIONAL ATOMIC ENERGY AGENCY, Site Evaluation for Nuclear Installations, IAEA Safety Standards Series No. SSR-1, IAEA, Vienna (2019).

[6]   RADIATION AND NUCLEAR SAFETY AUTHORITY, Safety design of a nuclear power plant, YVL B.1, STUK, Helsinki (2019), https://www.stuklex.fi/en/ohje/YVLB-1.

[7]   INTERNATIONAL ATOMIC ENERGY AGENCY, IAEA Safety Glossary: Terminology Used in Nuclear Safety and Radiation Protection, 2018 Edition, IAEA, Vienna (2019).

[8]   INTERNATIONAL ATOMIC ENERGY AGENCY, Design Provisions for Withstanding Station Blackout at Nuclear Power Plants, IAEA-TECDOC-1770, IAEA, Vienna (2015).

[9]   INTERNATIONAL ATOMIC ENERGY AGENCY, Considerations on the Application of the IAEA Safety Requirements for the Design of Nuclear Power Plants, IAEA-TECDOC-1791, IAEA, Vienna (2016).

[10]   OFFICE FOR NUCLEAR REGULATION, Safety assessment principles for nuclear facilities, 2014 Edition, Revision 1, ONR, Bootle, UK (2020).

[11]   GENERAL ELECTRICS HITACHI NUCLEAR ENERGY, The ABWR Plant General Description (2007), https://nuclear.gepower.com/content/dam/gepower-nuclear/global/en_US/documents/ABWR%20General%20Description%20Book.pdf.

[12]   GENERAL ELECTRICS HITACHI NUCLEAR ENERGY, ABWR Nuclear Power Plant (2007), https://nuclear.gepower.com/build-a-plant/products/nuclear-power-plants-overview/abwr.

[13]   INTERNATIONAL ATOMIC ENERGY AGENCY, Deterministic Safety Analysis for Nuclear Power Plants, IAEA Safety Standards Series No. SSG-2 (Rev. 1), IAEA, Vienna (2019).

[14]   INTERNATIONAL ATOMIC ENERGY AGENCY, Design of the Reactor Containment and Associated Systems for Nuclear Power Plants, IAEA Safety Standards Series No. SSG-53, IAEA, Vienna (2019).

[15]   INTERNATIONAL ATOMIC ENERGY AGENCY, Current Approaches in Member States to Analysis of Design Extension Conditions with Core Melting for New Nuclear Power Plants, IAEA-TECDOC-1982, IAEA, Vienna (2021).

[16]   INTERNATIONAL ATOMIC ENERGY AGENCY, Design of Electrical Power Systems for Nuclear Power Plants, IAEA Safety Standards Series No. SSG-34, IAEA, Vienna (2016).

[17]   INTERNATIONAL ATOMIC ENERGY AGENCY, Design of Instrumentation and Control Systems for Nuclear Power Plants, IAEA Safety Standards Series No. SSG-39, IAEA, Vienna (2016).

[18]   INTERNATIONAL ATOMIC ENERGY AGENCY, Design of the Reactor Core for Nuclear Power Plants, IAEA Safety Standards Series No. SSG-52, IAEA, Vienna (2019).

[19]   INTERNATIONAL ATOMIC ENERGY AGENCY, Design of the Reactor Coolant System and Associated Systems for Nuclear Power Plants, IAEA Safety Standards Series No. SSG-56, IAEA, Vienna (2020).

[20]   EUROPEAN ATOMIC ENERGY COMMUNITY, FOOD AND AGRICULTURE ORGANIZATION OF THE UNITED NATIONS, INTERNATIONAL ATOMIC ENERGY AGENCY, INTERNATIONAL LABOUR ORGANIZATION, INTERNATIONAL MARITIME ORGANIZATION, OECD NUCLEAR ENERGY AGENCY, PAN AMERICAN HEALTH ORGANIZATION, UNITED NATIONS ENVIRONMENT PROGRAMME, WORLD HEALTH ORGANIZATION, Fundamental Safety Principles, IAEA Safety Standards Series No. SF-1, IAEA, Vienna (2006).

[21]   RADIATION AND NUCLEAR SAFETY AUTHORITY, Regulation on the Safety of a Nuclear Power Plant, Y/1/2018, STUK, Helsinki (2018), https://www.stuklex.fi/en/maarays/stuk-y-1-2018.

[22]   INTERNATIONAL ELECTROTECHNICAL COMMISSION, Nuclear power plants – Instrumentation and control systems important to safety – Requirements for coping with common cause failure (CCF), IEC 62340, IEC, Geneva (2007).

[23]   INTERNATIONAL ATOMIC ENERGY AGENCY, Criteria for Diverse Actuation Systems for Nuclear Power Plants, IAEA-TECDOC-1848, IAEA, Vienna (2018).

[24]   MULTINATIONAL DESIGN EVALUATION PROGRAMME, Common position on the treatment of common cause failure caused by software within digital safety systems, MDEP Generic Common Position No DICWG-01, OECD-NEA, Paris (2013).

[25]   INTERNATIONAL ATOMIC ENERGY AGENCY, Assessment of Defence in Depth for Nuclear Power Plants, Safety Reports Series No. 46, IAEA, Vienna (2005) (under revision).

[26]   WESTERN EUROPEAN NUCLEAR REGULATORS' ASSOCIATION, Safety Reference Levels for Existing Reactors – Update in relation to lessons learned from TEPCO Fukushima Dai-ichi accident, WENRA Reactor Harmonisation Working Group (2014).

[27]   OECD NUCLEAR ENERGY AGENCY, Implementation of Defence in Depth at Nuclear Power Plants. Lessons Learnt from the Fukushima Daiichi Accident, NEA Report No. 7248, OECD-NEA, Paris (2016).

[28]   WESTERN EUROPEAN NUCLEAR REGULATORS' ASSOCIATION, Safety of New NPP Designs, WENRA Reactor Harmonisation Working Group (2013).

[29]   KOREA INSTITUTE OF NUCLEAR SAFETY, Regulatory standards and guidelines for PWR in Korea, Chapter 16: Severe accident and risk assessment, KINS/RS-N16.00 (Rev. 2) (2016).

[30]   NUCLEAR SAFETY AND SECURITY COMMISSION OF KOREA, Regulation on the Scope of Accident Management and the Detailed Criteria for Evaluating Accident Management Capabilities, Notice 2017-34 (2017).

[31]   OECD NUCLEAR ENERGY AGENCY, Status Report on Hydrogen Management and Related Computer Codes, NEA/CSNI/R(2014)8, OECD-NEA, Paris (2014).

[32]   OECD NUCLEAR ENERGY AGENCY, HYMERES project summary report: Resolving Complex Safety Relevant Issues Related to Hydrogen Release in Nuclear Power Plant Containments During a Postulated Severe Accident, NEA/CSNI/R(2018)11, OECD-NEA, Paris (2018).

[33]   EUROPEAN NUCLEAR SAFETY REGULATORS GROUP, EU "Stress Tests" Specifications: Annex 1 of the declaration of ENSREG dated 13 May (2011).

[34]   EUROPEAN NUCLEAR SAFETY REGULATORS GROUP, Peer review report, Stress tests performed on European nuclear power plants, ENSREG, Brussels (2012).

[35]   OECD NUCLEAR ENERGY AGENCY, Safety Margin Action Plan (SMAP) Final Report, NEA/CSNI/R(2007)9, OECD-NEA, Paris (2007).

[36]   OECD NUCLEAR ENERGY AGENCY, Status Report on Ex-Vessel Steam Explosion, NEA/CSNI/R(2017)15, OECD-NEA, Paris (2017).

[37]   OECD NUCLEAR ENERGY AGENCY, State-of-the-Art Report on Molten Corium Concrete Interaction and Ex-Vessel Molten Core Coolability, NEA/CSNI/R(2016)15, OECD-NEA, Paris (2016).

[38]   OECD NUCLEAR ENERGY AGENCY, Status report on Hydrogen management and related computer codes, NEA/CSNI/R(2014)8, OECD-NEA, Paris (2014).

[39]   NUCLEAR REGULATORY COMMISSION, Evaluation of Severe Accident Risks: Grand Gulf, Unit 1, NUREG/CR-4551, SAND86-1309, Vol. 6, Rev. 1, Part 1 (1990).

[40]   NUCLEAR REGULATORY COMMISSION, Evaluation of Severe Accident Risks: Peach Bottom, Unit 2, NUREG/CR-4551, SAND86-1309, Vol. 4, Rev. 1, Part 1 (1990).

[41]   NUCLEAR REGULATORY COMMISSION, Individual Plant Examination: Submittal Guidance, NUREG-1335 (1989).

[42]   NUCLEAR REGULATORY COMMISSION, State-of-the-Art Reactor Consequence Analyses Project: Uncertainty Analysis of the Unmitigated Long-Term Station Blackout of the Peach Bottom Atomic Power Station, NUREG/CR-7155, SAND2012-10702P (2016).

[43]   ELECTRIC POWER RESEARCH INSTITUTE, Advanced Nuclear Technology: Advanced Light Water Reactor Utility Requirements Document, Revision 13, Technical Report No. 3002003129, EPRI, Palo Alto, CA (2014).

[44]   ELECTRIC POWER RESEARCH INSTITUTE, Modular Accident Analysis Program 5 (MAAP5) Applications Guidance: Desktop Reference for Using MAAP5 Software – Phase 3 Report, Technical Report No. 3002010658, EPRI, Palo Alto, CA (2017).

[45]   INTERNATIONAL ATOMIC ENERGY AGENCY, Design of Nuclear Installations Against External Events Excluding Earthquakes, IAEA Safety Standards Series No. SSG-68, IAEA, Vienna (2021).

[46]   INTERNATIONAL ATOMIC ENERGY AGENCY, Seismic Design for Nuclear Installations, IAEA Safety Standards Series No. SSG-67, IAEA, Vienna (2021).

[47]   FEDERAL ENVIRONMENTAL, INDUSTRIAL AND NUCLEAR SUPERVISION SERVICE (ROSTECHNADZOR), Record of external natural and human-induced impacts on nuclear facilities (NP-064-17), No. 514, Moscow (2017).

[48]   FEDERAL ENVIRONMENTAL, INDUSTRIAL AND NUCLEAR SUPERVISION SERVICE (ROSTECHNADZOR), Basic Recommendations to the Preparation of the Probabilistic Safety Analysis (Level 1 PSA) for NPP Power Unit in Case of Initiating Events Caused by Impacts of Natural and Man-Induced Origin (RB-021-14), No. 396, Moscow (2014).

[49] NATIONAL NUCLEAR SAFETY ADMINISTRATION, Site Survey for Nuclear Power Plant, Chinese National regulatory guide HAD 101/07 (2007).

[50] NATIONAL NUCLEAR SAFETY ADMINISTRATION, External Human Induced Events in Relation to Nuclear Power Plant Siting, Chinese National regulatory guide HAD 101/04-1989 (1989).

[51] NATIONAL NUCLEAR SAFETY ADMINISTRATION, Design Basis Tropical Cyclone for Nuclear Power Plants, Chinese National regulatory guide HAD 101/11-1991 (1991).

[52] NUCLEAR REGULATION AUTHORITY, Ordinance Prescribing Standards for the Location, Structures, and Equipment of Commercial Power Reactors and their Auxiliary Facilities, Reactor Establishment Permit Ordinance (2013), https://www.nsr.go.jp/english/regulatory/index.html.

[53] THE EUROPEAN UNION PER REGULATION, Eurocode – Basis of structural design, The European Union Per Regulation 305/2011, Directive 98/34/EC, Directive 2004/18/EC, EN 1990 (2002).

[54] EUROPEAN UTILITY REQUIREMENTS ORGANISATION, European Utility Requirements for LWR Nuclear Power Plants, Revision E (2017).

[55] NUCLEAR REGULATORY COMMISSION, Office of Nuclear Regulatory Research, Ultimate Heat Sink for Nuclear Power Plants, Regulatory Guide 1.27, Revision 3, Washington, DC (2015).

[56] CANADIAN NUCLEAR SAFETY COMMISSION, Design of Small Reactor Facilities, RD-367, CNSC, Ottawa (2011).

[57] CANADIAN NUCLEAR SAFETY COMMISSION, Design of Reactor Facilities: Nuclear Power Plants, REGDOC-2.5.2, CNSC, Ottawa (2014).

[58] AUTORITE DE SÛRETÉ NUCLÉAIRE, Guide de l'ASN n° 22. Conception des réacteurs à eau sous pression (2017).

[59] INTERNATIONAL ATOMIC ENERGY AGENCY, Accident Management Programmes for Nuclear Power Plants, IAEA Safety Standards Series No. SSG-54, IAEA, Vienna (2019).

[60] NUCLEAR REGULATORY COMMISSION, Flexible mitigation strategies for beyond-design-basis events, Regulatory Guide 1.226, Revision 0, Washington, DC (2019).

[61] NUCLEAR ENERGY INSTITUTE, Diverse and flexible coping strategies (FLEX) implementation guide, NEI 12-06, Revision 4, NEI, Washington, DC (2016).

[62] AUTORITE DE SÛRETÉ NUCLÉAIRE, Technical guidelines for the design and construction of the next generation of Nuclear Power Plants with Pressurized Water Reactors. Adopted during the GPR/German experts plenary meetings held on October 19th and 26th (2000).

[63] THE AMERICAN SOCIETY OF MECHANICAL ENGINEERS, Section III, Rules for construction of nuclear facility components, ASME Boiler and Pressure Vessel Code Division 1, Subsection NB: Class 1 components, ASME BPVC-III NE (2021).

Annex I

CHINA, CNNC'S HPR1000

I-1. GENERAL DESIGN FEATURES OF HPR1000

I-1.1. Background

HPR1000 is an evolutionary advanced PWR developed by China National Nuclear Corporation (CNNC). The design utilizes proven technology based on design, construction and operating experience of a large PWR fleet in China, and incorporates a series of advanced design features to meet the utility requirements for advanced light water reactors and to address the latest nuclear safety requirements, including the safety issues relevant to the Fukushima Daiichi accident.

The R&D of HPR1000 can be traced back to 1999, and went through three phases, which are represented by three successively developed PWR models, known as CNP1000, CP1000 and ACP1000 (officially designated as HPR1000 in 2013), respectively. After the Fukushima Daiichi accident, the programme was gathering pace due to the preference of the authority towards advanced PWRs with higher safety performance, with complementary research induced by the feedback from the accident. In 2015, the Chinese government approved the construction of HPR1000 demonstration project, Fuqing NPP Units 5&6. As of May 2022, four HPR1000 units have been connected to the grid, including two domestic units (Fuqing Units 5&6) and two units abroad (Pakistan Karachi Units 2&3), while four HPR1000 units are under construction (as of July 2022).

I-1.2. Design features of HPR1000

As an active and passive advanced PWR, HPR1000 adopts proven technology and integrates the feedbacks of the Fukushima Daiichi accident.

The reactor core of HPR1000 is loaded with 177 fuel assemblies, ensuring sufficient thermal margin while increasing output power. With innovative zircaloy cladding material and grid, nozzle and guide tube design, the fuel subassembly has an 18-month refuelling cycle.

The application of active and passive safety design philosophy is an important innovation for HPR1000. The design inherits the mature and reliable active technology (e.g. the engineered safety systems including SIS, AFS and CSS, validated by long term engineering practice from existing NPPs), and introduces passive systems (e.g. PRS, PCS, active and passive cavity injection and cooling system) as a backup for active systems in case of loss of AC power. More specifically, both active and passive features are employed to guarantee the safety functions of emergency core cooling, residual heat removal, in-vessel retention of molten core, and containment heat removal. By introducing passive systems, the accident mitigation capacity and the residual heat removal capacity have been enhanced under DECs, successfully implementing DiD.

The design of HPR1000 also incorporates comprehensive severe accident preventive and mitigating measures against various severe accident phenomena, including HPME, hydrogen detonation, basement melt-through and long term containment overpressure. The PSA result of

Fuqing Units 5&6 shows that the CDF is less than 10^{-6}/r.y, and that the LRF is less than 10^{-7}/r.y.

Furthermore, new features were implemented on HPR1000 after the Fukushima Daiichi accident, for coping with such accident scenario, such as:

- Emergency power and cooling water supply solution;
- Improved monitoring and cooling capability of the SFP;
- Enhanced habitability and availability of emergency features.

The capability against extremely external events has also been effectively enhanced, by adopting the seismic input of $0.3g$ peak ground acceleration and designing and implementing protection features against large commercial aircraft crash.

The operational performance of HPR1000 is also in line with the requirements set for third generation PWRs, e.g. the overall plant availability goal is greater than 90% considering all forced and planned outages, 30 minute non-intervention of operator, 72 hours autonomy, plant design lifetime of 60 years, and a 18 month refuelling cycle.

I-2. DEC AND 'PRACTICAL ELIMINATION'

I-2.1. Features for design extension conditions without significant fuel degradation (DEC-A)

In the design of HPR1000, a set of DECs are derived based on engineering judgement, deterministic and probabilistic method evaluation, to further improve the plant safety.

In the identification of DECs without significant fuel degradation (DEC-A) for HPR1000, PSA methods and models are used to identify and determine extremely unlikely events and multiple failure events, while also considering deterministic analyses and engineering judgement.

In the process of applying the PSA model to the identification of DEC-A conditions, considering the large number of potential sequences, the sequences are selected and truncated according to the sequence frequency. Only the sequences that might have an important impact on safety are considered and analysed as initial DEC-A. The determination of the DEC-A frequency cut-off value is directly related to the safety goals of the plant. DEC-A safety features, such as the SBO power supply system, the PHRS on the secondary side, and the diverse actuation system, are removed from the PSA model when identifying DEC-A conditions, consistently with the purpose of DEC-A analyses.

In response to the complex accident sequence of DEC-A, measures such as the SBO power supply system, the diversified cooling system of the safety injection pump, the PHRS on the secondary side, and the PCS have been included in the design. All safety features for DECs can effectively mitigate the consequences of the accident, enhance the ability of NPPs to deal with DECs that are more serious than DBAs, effectively avoid unacceptable radiological consequences, and significantly improve the safety of NPPs.

Through an iterative design process, it is ensured that the selection of DEC-A sequences and the configuration of DEC-A safety features can meet the safety goals for the NPP.

The analysis of DECs without significant fuel degradation adopts either best estimate or conservative methods.

Nominal values of the initial operating conditions of DEC-A accident analysis are considered. The conservative assumptions of the initial state in the DBA analysis can also be used in the design extension condition analysis. When carrying out uncertainty calculation or sensitivity analysis, specific parameters are selected, and deviations are determined.

The general principle of DEC-A accident evaluation is that only the system equipment available under DECs can be used for the DEC analysis. The analysis carried out should include the identification of features used or capable of preventing and mitigating DECs. These features should be independent from the features used in the more frequent accidents as far as practicable and should be able to perform the expected function under the environmental conditions corresponding to DEC-A, and its reliability should be consistent with the functions required to be realized.

Unless it is assumed to be unavailable in the multiple failure events, the system is considered to be available if its operation does not exceed its design range. The SFC is not adopted.

I-2.2.   Features for design extension conditions with core melting (DEC-B)

The DECs with core melting of HPR1000 are identified based on the consideration of a combination of severe accident phenomena that can lead to containment failure and large radioactive release with a non-negligible likelihood.

The following severe accident phenomena are identified on the basis of a combination of deterministic analysis, probabilistic analysis and engineering judgement:

- HPME and DCH;
- Ex-vessel steam explosion;
- MCCI;
- Combustion or explosion of combustible gas in the containment;
- Overpressurization of the containment.

For each of the above severe accident phenomena, a list of scenarios is selected to generate bounding or representative physical parameters for the design of the corresponding safety features, as shown in Table I-1.

TABLE I-1. DEC WITH CORE MELTING (DEC-B) AND CORRESPONDING PHENOMENA AND SAFETY FEATURES

No.	Phenomena	DEC-B conditions	Dedicated DEC-B safety feature
1	High-pressure melt ejection (HPME) and direct containment heating (DCH)	Loss of feedwater accident together with failure of PRS and active safety injection	Dedicated RCS depressurization system
2	Ex-vessel steam explosion and molten core concrete interaction (MCCI)	Large LOCA together with failure of active safety injection	Cavity injection and cooling system
3		Medium LOCA together with failure of active safety injection	
4		Small LOCA together with failure of active safety injection	
5		SBO together with failure of secondary heat removal	
6	Combustion or explosion of combustible gas in containment	Large LOCA together with failure of active safety injection	Containment hydrogen combination system
7		Medium LOCA together with failure of active safety injection	
8		Small LOCA together with failure of active safety injection	
9		SBO together with failure of secondary heat removal	
10	Overpressurization of containment	Large LOCA together with failure of active safety injection and containment spray	Passive containment heat removal system (PCS)
11		MSLB in containment together with failure of auxiliary feedwater supply, active safety injection and containment spray	

In order to demonstrate the achievement of a controlled and/or a safe state for a severe accident, a series of acceptance criteria, including both radioactive acceptance criteria and technical (i.e. thermal-hydraulic parameters related) acceptance criteria, are first defined. A deterministic severe accident analysis is then conducted for the above DEC-B accident scenarios to demonstrate that such predefined acceptance criteria are met.

I-2.3. 'Practical elimination'

There were wide and continuous discussions about the issues of 'practical elimination' in China, and consensus is starting to emerge. The viewpoints on this issue, included in the document of

Nuclear Safety Review Principles for HPR1000 drafted by the National Nuclear Safety Administration, are presented as follows:

- The concept of 'practical elimination' is only applied to accident condition or accident sequence.
- For the accident conditions or accident sequences that might cause early or large radioactivity releases, reliable design provisions shall be incorporated to achieve the target of 'practical elimination'.
- The selection of accident conditions or accident sequences is based on a judgement on their frequency, and it is recommended to adopt the value of 10^{-7}/r.y as a complementary judgement for 'practical elimination'.
- With regards to the capability to withstand external natural hazards, it is required to evaluate the margin to protect the items which are necessary to prevent early or large radioactivity releases and to avoid cliff edge effects.

According to the above viewpoints, 'practical elimination' should be considered as the outcome of DiD. More explicitly, the accident conditions or accident sequences that might cause early or large radioactivity releases are 'practically eliminated' because of the design provisions in levels 1 to 4.

The detailed demonstration of HPR1000 'practical elimination' is shown in the Appendix. By adopting a set of reliable and effective mitigation measures, the event sequences that would lead to an early radioactive release or a large radioactive release can be 'practically eliminated' for HPR1000.

I-3. DEFFENCE IN DEPTH FEATURES AND INDEPENDENCE

I-3.1. HPR1000 defence in depth features

As an advanced PWR, HPR1000 uses both active and passive measures to provide effective multilevel defence, which maximizes the balance between different DiD levels. These comprehensive active and passive measures can guarantee the fulfilment of fundamental safety functions under accident conditions and ensure the effectiveness of physical barriers.

In fact, the design of HPR1000 provides a practical scheme that combines active and passive measures, gives full play to the advantages and characteristics of both active and passive means, and at the same time ensures that it can meet DiD requirements:

- In general, for DBAs (DiD level 3a), active means are mainly adopted, including two trains of SIS, AFS and CSS;
- When dealing with DECs (DiD level 3b and level 4), additional passive safety measures, including the PRS, the PCS, the containment hydrogen recombination system and the passive cavity injection system, are set specifically to cope with the multiple failure of active safety systems.

This design greatly reduces the possibility of common course failure across different DiD levels, avoids excessively and unilaterally strengthening the defence capability against DBAs, and emphasizes the design balance of each DiD level.

The major design measures for the DiD levels of HPR1000 accident conditions are shown in Table I-2. The use of active and passive safety measures can provide effective multilevel defence and achieve the balance between the defence capability of the various levels. Specifically, passive systems can provide various diversity in mitigating DECs.

TABLE I-2. MAJOR SAFETY FEATURES AT DID LEVELS 3 AND 4 FOR HPR1000

DID level	Condition	Major safety features
3a	DBA (with postulated single initiating event)	• Safety injection system (SIS) (including accumulator[a]) • Containment spray system (CSS) • Auxiliary feedwater system • Turbine bypass system • Emergency diesel generators (EDGs)
3b	DEC without significant fuel degradation (multiple failure sequences)	• Passive residual heat removal system from secondary side[a] • Passive containment heat removal system (PCS)[a] • Emergency boron injection system • Diversified cooling chain and heat sink • Diverse actuation system • Standstill sealing of main pump[a] and emergency sealing water injection • SBO DG
4	DEC with core melting	• Fast depressurization system of primary loop[a] • Active and passive* cavity injection system • Passive containment heat removal system[a] • Containment hydrogen recombination system* • Containment filtration and venting system • High-point venting system of reactor pressure vessel (RPV) • 72-hour batteries • Inhabitability system of main control room

[a] Passive feature.

I-3.2. Independence of defence in depth

It is a common sense that achieving full independence between different DiD levels is not practical. Considering some practical limitations, the independence of DiD has been implemented 'as far as practicable' in the HPR1000 design. This section will illustrate the considerations and design for the independence between levels 3a and 3b, as well as between levels 3a and 4.

The safety system of level 3a will still be used in part of the DEC-A sequences mitigation, so the levels 3a and 3b cannot be completely independent. Specific considerations for independence between levels 3a and 3b are as follows:

- The diverse actuation system, as backup in case of multiple failure of the reactor protection system, adopts a different platform from the protection system and achieves maximum physical and electrical isolation.
- As for ATWS, the emergency boron injection system adopts a shutdown mode which is different from the control rods, and these two shutdown modes are independent of each other through triggering separated ATWS modules.
- As a backup means of heat removal through the secondary side, the passive residual heat removal system from the secondary side does not rely on the power supply or steam source required for the operation of the AFS, but only needs to open the isolation valves and establish a natural heat removal circulation.

- As a backup means of containment heat removal after accidents, the PCS does not have any interface components with the CSS. By opening the isolation valves and establishing a natural circulation, the containment heat can be evacuated.
- The SBO power supply system and the EDGs are arranged in different buildings. The support system and auxiliary system required for the operation of SBO DGs are also independently set up, and independent distribution boards are also adopted.
- The cooling water trains for components necessary to mitigate accidents and for normal operation components are set separately. For example, the electrical building chilled water system set an independent air cooled chiller, equipped with SBO power, for safety injection pumps cooling and ensure the safe operation of the control room, control cabinet room during SBO condition. The cooling sources are different between air cooled chiller and water cooled chiller, and the layout of these chillers is also physically isolated.

The independence between levels 3a and 4 has been implemented at a sufficient level for HPR1000, as follows:

- For the fast depressurization system of the primary loop designed to eliminate the risk of HPME, the two sets of quick relief valves are independent of the three sets of pressurizer safety valves that perform overpressure protection and are not required to be used under AOOs or DBAs. The quick relief valve is an electric gate valve, and the opening mechanism is different from the pilot operated safety valve.
- Two safety-classified passive recombiners of the containment hydrogen combination system are designed for DBAs, while 31 separate passive recombiners (completely independent from the previous ones) are designed to cope with severe accidents.
- The PCS (only required for individual DEC-A sequences such as LOCA plus safety spray failure) for containment heat removal after severe accidents is independent from the CSS.
- The cavity injection and cooling system is dedicated to implement in-vessel retention strategy during severe accidents.
- The severe accidents, the I&C system is independent from the reactor protection system and the diverse actuation system, with separate cabinets.
- Two 72-hour uninterrupted power supply systems are set independently for the instrument and control devices and valves which are necessary to mitigate severe accidents, with separate DC and uninterrupted power supply switchboards.

I-4. EXTERNAL HAZARDS

I-4.1. Hazards list

A comprehensive list of potential external hazards (as well as the combination sets) was identified according to the Chinese national guide HAD, IAEA Safety Guides and other requirements as reference, based on DSA combined with PSA methods. The effects of hazards are fully evaluated with conservative protection measures adopted [I-1].

a) Typical external natural hazards considered in the design of NPPs include: earthquake; external flood; extreme wind; precipitation, snow, icing; drought; lighting; external fire (from natural phenomenon); biological phenomena (such as water intakes blocked by aquatic organisms).
b) Typical external human induced hazards considered in the design of NPPs include: aircraft crash; hazards from industrial or traffic environment near the site (such as external explosion, external missiles, hazardous gas cloud); electromagnetic interference;

sabotage (such as collision of ships with accessible safety structures, or collision of vehicles with SSCs).

c) Credible combinations of hazards are identified as follows:
  i. Combination between external hazard and its consequential hazard, e.g.:
     - Earthquake and internal fire/flood/pipe failure/heavy loads;
     - Earthquake and external flood/explosion;
     - LOOP resulting from earthquake or flood or extreme wind or tornado or lighting or explosion or aircraft crash;
     - Aircraft crash and explosion;
  ii. Combination between hazards as the result of a common initiating event, e.g. precipitation and flood/extreme wind/lighting, extreme wind and flood;
  iii. Combination between external hazard and independent external or internal hazard identified either by probability analysis or considering for design margin based on engineering judgement, e.g. earthquake and DBA, biological phenomena/extreme temperature and loss of UHS, flood/extreme wind and biological phenomena.

I-4.2. Design process for protecting against external hazards

For the identified hazards, the effects are fully evaluated, corresponding design basis and loads are set according to guides, standards and requirements of protection, design principles and specified design loads, and suitable and conservative protection measures (such as source appropriate layout, physical separation, capability of items to withstand or be protected against a hazard, redundancy, diversity) are adopted, guided by the DiD philosophy, to ensure that the plant safety will not be impaired by internal or external hazards.

The design and analysis steps are illustrated in Figs I-1 and I-2.

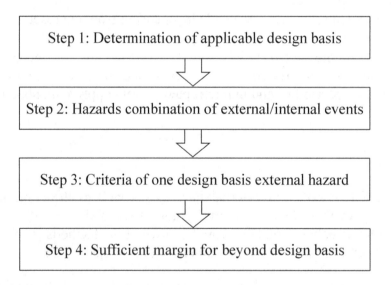

FIG. I-1. Approach for designing against external hazards.

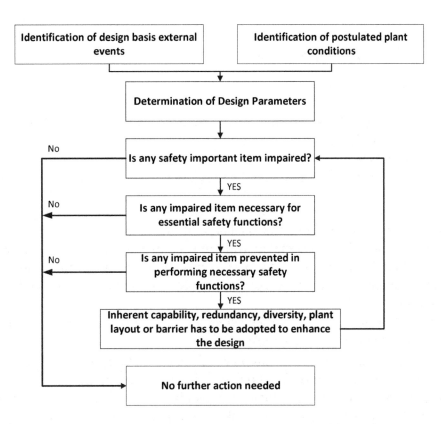

FIG. I-2. Process for designing against external hazards.

I-4.3. Hazard protection example: flood

I-4.3.1.   Determination of the design basis

(a)   Hydrological hazards

Hydrological phenomena that are generated by relevant amount of water and which may cause flooding or low water conditions should be considered. Relevant amount of water are all oceans, seas, estuaries, lakes, reservoirs, rivers and canals that may produce flooding on or adjacent to the NPP site.

The most important phenomena include the following: storm surges; waves; tsunamis; seiches; extreme precipitation; sudden releases of water from natural or artificial storage.

Other hydrological phenomena that could cause hazards to the installation include the following:

- Water level rising upstream or falling downstream caused by e.g. obstruction of a river channel by landslides or by jams caused by ice, logs, debris or volcanic materials;
- Landslides or avalanches into water bodies;
- Waterspouts;
- Deterioration or failure of facilities on the site or near site facilities (e.g. canals, water retaining structures or pipes);
- Swelling of water in a channel due to a sudden change in the flow rate; the origin may be natural (for example a tidal bore) or artificial (as in the case of closure of a hydroelectric plant);

- Variation of groundwater levels;
- Subsurface freezing of supercooled water (frazil ice).

(b)   Combinations of events

A suitable combination of flood causing events depends on the specific characteristics of the site and involves considerable engineering judgement. The following are examples of combinations of events that cause floods for use in determining the design conditions for flood defence. The design basis flood should be the maximum of the following load combinations:

1.   For an estuary site where the following items are of importance: the astronomical tide; the storm surge; wave runup; the discharge of the river;
2.   The design basis flood associated with an established probability of exceedance (e.g. 10^{-4}) for the following combination of events should be determined (including several statistical parameters, where some of them have a strong correlation and others have no correlation):
      - High water level (which is a function of astronomical high water, storm surge (wind) and river discharge);
      - Cumulated with wave runup (which is a function of water level, wave height, wave period (wind) and geometry of the construction).
3.   According to the collection of historical data in a given country, this evaluation can be performed in a conservative way, taking the maximum value among the following proposed load combinations (designated as A, B, C and D):
      A) **Combination A:**
      - Design water level (based on the spring tide, a value of 10^{-4} storm surge value on the coast, and the average value for the river discharge), combined with the item below;
      - Wave runup (with the most probable wave height and wave period, with due account taken of the geometry of the construction).
         The wave parameters can be derived through a wave model using the design water level and the same wind as used for the calculation of the design water level with a hydraulic model.
      B) **Combination B:**
      - High water level (based on the spring tide, a value of 10^{-2} storm surge value on the coast, and 10^{-1} for the river discharge), combined with the item below;
      - Wave runup (with the most probable wave height and wave period, with due account taken of the geometry of the construction).
         The probability of coincidence of the storm surge with the river flood has been taken (conservatively) as 10^{-1}.
      C) **Combination C:**
      - High water level (based on the spring tide, a value of 10^{-1} storm surge value on the coast, and 10^{-2} for the river discharge), combined with the item below;
      - Wave runup (with the most probable wave height and wave period, with due account taken of the geometry of the construction).

D) **Combination D:**
- High water level (based on the spring tide, no storm surge value on the coast, and 10^{-4} for the river discharge), combined with the item below;
- 0.5 m freeboard.

I-4.3.2.   Determination of beyond design basis

Based on the feedback from the Fukushima Daiichi accident, the beyond design basis for flood is set by the combination of the design basis flood and the precipitation on a 1000-year return period.

I-4.3.3.   Design of flood protection

Protections against design basis flood and beyond design basis flood for an NPP include the following:

(i) Design of flood protections for the site, with the options below:
- The 'dry site' concept;
- Permanent external barriers such as levees, sea walls and bulkheads.
(ii) Design of a drainage system or facility for the site;
(iii)Flood protection of items important to safety;
(iv)Protection for intake and drainage system and protective structure.

REFERENCES TO ANNEX I

[I-1] CHINA NATIONAL NUCLEAR CORPORATION, Design and Analysis Criteria of Protection against External Hazards for Pressurized Water Reactor Nuclear Power Plants, Q/CNNC HLBZ CA 1 (2018).

Annex II

FRANCE

II-1. THE EUROPEAN PRESSURIZED WATER REACTOR (EPR)

II-1.1. Historical technical basis of the EPR

The EPR, jointly developed by the French and German industries, is an evolutionary development of PWRs currently operating in Germany and France, where the initial construction was planned, hence on a European basis.

A basis for the design was provided by the definition of fundamental safety requirements for future PWRs by the French 'Groupe Permanent Réacteur' and the German Commission for Reactor Safety: the Technical Guidelines [II-1] adopted during the plenary meetings held on October 19th and 26th 2000. In addition to the Technical Guidelines [II-1], the EPR development considered international standards, such as IAEA Safety Standards Series No. NS-R-1 (eventually superseded by SSR-2/1 (Rev. 1) [II-2]).

The French regulatory practice regarding the design of new NPPs is now detailed in the 2017 Autorité de Sûreté Nucléaire (ASN) guide n° 22 [II-3], accounting for SSR-2/1 (Rev. 1) [II-2] (see further details in Section II-2).

The following description is mainly based on the Flamanville Unit 3 (FA3 EPR), a relevant practical illustration of the recent French practice, considering as far as possible SSR-2/1 (Rev. 1) [II-2] and the ASN guide n° 22, both issued in parallel to the FA3 EPR design development.

II-1.2. Design features and rationale

From the very beginning, the aim of the EPR project was to achieve an enhanced safety level in comparison to the existing French and German NPPs. Additional information is available in IAEA's ARIS database [II-4].

II-1.2.1. Main safety principles and consideration of defence in depth

The EPR design is part of the generation III+ PWRs developed by AREVA and Mitsubishi Heavy Industries, respectively, the EPR and the Advanced PWR.

The DiD concept is fully applied to the design. This was initially derived in compliance with Section A1.2 of the Technical Guidelines [II-1]. The implementation of the DiD concept to the design is ensured by a series of five levels to prevent accidents and to provide an appropriate protection should the prevention fail:

1. Level 1: A combination of conservative design, quality assurance, and surveillance activities is applied to prevent deviations from normal operation (failure of equipment or control loop).
2. Level 2: Detection of deviations from normal operation as well as protection devices (as well as control systems provided to cope with them) are implemented in the design. This level aims at preserving the integrity of the first barrier (the fuel cladding) and the second barrier (the reactor coolant pressure boundary).

3. Level 3: Engineered safety features (ESFs) and protective systems, as well as operating procedures, are provided to mitigate accidents, if they were to occur, to confine the radioactive releases, if any, and consequently to prevent their development into severe accidents.
4. Level 4: Measures are implemented to preserve the integrity of the containment and enable mitigation of potential severe accidents.
5. Level 5: Off-site emergency response.

The third level is split into two sublevels:

- Level 3a for design basis accident (DBC-3 and DBC-4);
- Level 3b for complex sequences (DECs without significant fuel degradation, indicated here as DEC-A).

In addition to the verification that the level 2 control and safety systems prevent the occurrence of an accident, the design of the FA3 EPR is verified to ensure that sufficient independence is implemented. In practice, the verification is to ensure that, when the failure of a safety system from the second level of DiD initiates an accident of the third level, the safety systems and safety features required to prevent the occurrence of a severe accident condition are independent from the failed system. Such situations of multiple failures are studied as part of DECs without significant fuel degradation (DEC-A).

II-1.2.2. Plant states considered in EPR design

The safety assessment approach adopted for the FA3 EPR design involves the analysis of potential internal events and their consequences. Practically and firstly, the design of the FA3 EPR considers a limited but comprehensive number of events, bounding the situations likely to occur as part of the plant operation and in any of the associated standard reactor states[1]. These single initiating events are grouped into different categories, based on reference to similar plants or conventional lists, their estimated frequency of occurrence and feedback from operating experience or engineering judgement. Four categories are identified as design basis conditions (DBCs):

- DBC-1: normal operating transients;
- DBC-2: anticipated operation occurrences (AOOs) or incidents that could happen more than once in the plant lifetime (frequency $f > 10^{-2}$/r.y);
- DBC-3: infrequent accidents or accidents with a low probability to occur during the plant lifetime (10^{-4}/r.y $< f < 10^{-2}$/r.y);
- DBC-4: hypothetical accidents or accidents not likely to happen during the plant lifetime (10^{-6}/r.y $< f < 10^{-4}$/r.y).

In the deterministic approach for DBCs, through the rules applied:

- A redundancy is required (SFC or Requirement 25 of SSR-2/1 (Rev. 1) [II-2]);

---

[1] These are the six standard reactor states for the FA3 EPR:
- State A: Power states and hot and intermediate shutdown states (close to hot shutdown).
- State B: Intermediate shutdown above 120°C, with heat removal in SG mode.
- State C: Intermediate and cold shutdown below 120°C, with heat removal in RHR mode.
- State D: Cold shutdown with reactor coolant pump (part of the RCS) open, and reactor cavity empty.
- State E: Cold shutdown with reactor coolant pump (part of the RCS) open, and reactor cavity flooded for refuelling.
- State F: Cold shutdown with the core totally unloaded.

- The redundancy is confirmed by the classification approach (Requirement 22 of SSR-2/1 (Rev. 1) [II-2]);
- A physical separation is required (Requirement 21 of SSR-2/1 (Rev. 1) [II-2]).

The FA3 EPR design is an evolutionary design benefitting from the experience of French and German NPPs in operation. As a main improvement, the FA3 EPR is considering the shutdown states as initial conditions for DBC analysis, as well as the analysis of DBC conditions for the SFP.

In order to further reduce the risk associated with the EPR, in a second step, two additional categories of DECs are considered:

- DEC-A: DECs without significant fuel degradation resulting from multiple failures (including CCF). The aim is to prevent significant fuel degradation thanks to specific design provisions;
- DEC-B: DECs with core melting.

II-1.2.3. Systems architecture

The fluid systems architecture is the result of an intensive exchange of information about design and operating experience between the EPR designers and the electric utilities involved in the design process. PSA was used at the very outset of the project for the definition of the following key principles of system architecture: redundancy, physical separation, and diversity.

Additional information is available in IAEA's ARIS database [II-4].

In practice, the safety systems and safety features are implemented in an overall four-train architecture allocated to four safety divisions. The physical separations among buildings are designed to prevent any hazard from spreading from one building to another. Two safety divisions are also housing the DEC safety features, with a minimum of separation from DBC safety systems.

II-1.2.4. Control of design extension conditions without significant fuel degradation (DEC-A)

In addition to DBAs, DEC-A sequences consider relevant and realistic combinations of single events. These DEC-A conditions are initially selected from previous reactor design and experience (i.e. ATWS, loss of UHS, SBO) and further completed with the support of the internal event level 1 PSA. The selected PSA sequences grouped into bounding DEC-A conditions are those likely to affect the overall probabilistic targets (the structured approach is based on several PSA targets). The DEC-A list is then completed with relevant situations of loss of electrical power, loss of UHS on a long term basis and some pipe leaks.

The DEC-A safety analyses of these bounding conditions are performed with the objective to demonstrate that a safe state (for the SFP studies, the objective is to avoid fuel uncovering) can be reached. This DEC-A final state is associated to the fulfilment of the following conditions:

- The core is subcritical;
- The residual heat is removed by primary or secondary systems;
- The radioactive releases remain tolerable (i.e. in accordance with predefined acceptance criteria).

The DEC-A safety features and the safety systems that may be credited in DEC-A condition studies are those not affected by the DEC-A conditions (e.g. where the DEC-A condition considers a CCF of the medium head safety injection pumps, the low head safety injection pumps may be credited). Independency of DEC-A features from DBC safety systems is considered sequence per sequence. A safety feature can be used in DBC and DEC-A provided that in each single sequence independency of DEC-A features is met.

The following rules are applied to the DEC-A safety analyses:

   i.   The SFC is not applied given the unlikelihood of the DEC-A conditions and the number of failures already considered;
   ii. The unavailability for preventive maintenance is not considered, given the unlikelihood of a DEC-A occurring precisely during the maintenance activity;
   iii. Operator actions are considered in accordance with emergency operating procedures:
        • A manual action from the main control room is taking place at least 30 minutes after the communication to the operator of the first significant information;
        • A local to plant action is taking place at least 60 minutes after the communication to the operator of the first significant information.

The FA3 EPR DEC-A acceptance criteria are the same as those considered for DBC-4 events, whereas the DEC-A rules to analyse DEC-A conditions are less conservative than those for DBC-4 events.

In order to reach the DEC-A acceptance criteria with a high confidence level (95%), the values for the dominant parameters in the safety analyses are taken at a reasonably bounding value, whereas other parameters (second order of significance parameters) may be taken as their best estimate values. The DEC-A safety analyses are separated from the PSA support studies, where best estimate assumptions may be used.

TABLE II-1: EXAMPLES OF ANALYSES OF DEC WITHOUT SIGNIFICANT FUEL DEGRADATION (DEC-A) FOR THE FA3 EPR

DEC-A sequence	Diversified strategy	Dedicated DEC-A feature
AOO + Failure of reactor trip (ATWS)	Boron injection by extra boration system (EBS)	ATWS signal (automatic actuation of EBS)
AOO + Failure of protection system (ATWS)	Diversified protection signals	ATWS diversified reactor trip
AOO + Failure of SGs cooling	Feed and bleed	Manual opening of pressurizer discharge system
LOOP + Failure of EDGs	Ultimate diesel generators	Manual start of ultimate diesel generators
SB-LOCA + CCF on medium head safety injection	Injection into RCS by low head safety injection	Manual actuation of fast secondary cool down
Total loss of heat sink – Hot shutdown	Heat removal by SGs	Manual resupply of emergency feedwater tanks
Total loss of heat sink – Cold shutdown (RHR system)	RCS boiling and containment cooling	Manual actuation of heat removal by the containment heat removal system (diversified cooling chain)

II-1.2.5. Control of severe accidents

As part of the main improvements included in the FA3 EPR design, from the initial design stages, severe accident conditions have been considered and mitigations means designed accordingly, in addition to the DBA safety systems and DEC-A safety features. All these measures make it possible to achieve the EPR DEC-B safety objectives.

Indeed, despite being very unlikely, as part of the FA3 EPR DECs with core melting (DEC-B), the occurrence of a severe accident is postulated. DEC-B conditions are identified with a deterministic approach based on the main parameters governing low-pressure severe accident condition. The DEC-B conditions studied as part of the safety analyses are sequences representative of the main physical challenge to the containment integrity: SB-LOCA and LB-LOCA, loss of AC power, total loss of feedwater.

The DEC-B conditions are used to define and verify the performances required for the safety features and safety systems, designed to mitigate DECs with core melting. The FA3 EPR DEC-B conditions and DEC-B safety features are defined and studied with the aim to demonstrate the achievement of the acceptance criteria such as:

- No emergency evacuation, except for the immediate vicinity to the NPP;
- Only limited (both in area and time) protective actions.

The DEC-B analyses are performed with the objective to demonstrate that a SASS can be reached. A SASS is defined as: the core melt is being cooled down, the decay heat is being removed, and the containment integrity is maintained.

The following rules apply:

- The SFC is not applied;
- The unavailability for preventive maintenance is not considered.

A SASS is reached and maintained by the following provisions:

- The control of hydrogen production by the containment PARs;
- The core melt stabilization within the containment by the core spreading area (otherwise called core catcher) and associated systems;
- The control of the containment pressure and temperature by the containment heat removal system;
- The limitation of radiological releases by the containment and associated systems (containment annulus, dynamic confinement of peripheral buildings, iodine filtration before release to the vent stack).

DEC-B safety features are qualified to severe accident conditions.

II-1.2.6. 'Practical elimination'

The deterministic demonstration of the avoidance of large radioactive releases or early radioactive releases which might result from severe accident conditions is provided by the severe accident safety analyses, including DEC-B and 'practically eliminated' conditions. These are complemented by a verification through level 2 PSA studies that uncontrolled and unacceptable (for people and the environment) releases are unlikely.

The FA3 EPR identified, through a systematic 'practical elimination' analysis based upon deterministic and probabilistic considerations, the following conditions to be 'practically eliminated', together with the necessary associated design provision, as presented in Table II-2.

TABLE II-2. DESIGN PROVISIONS IDENTIFIED FOR CONDITIONS TO BE 'PRACTICALLY ELIMINATED'

Condition to be 'practically eliminated'	Design provision
High-pressure core melt and direct containment heating (DCH)	Ultimate discharge valve
Fast uncontrolled increase of reactivity	Isolation of sources of dilution, automatically (isolation signal or use of non-return valves). Operational rules for enhanced isolation, control of boron concentration.
Steam explosions likely to lead to failure of the containment	Verification and justification that EPR severe accident design is fit for purpose, especially leaktightness of the area around the vessel ensuring the severe accident spreading area is dry by preventing water from entering this area of the reactor building.
Hydrogen combustion processes endangering the containment integrity	Verification and justification of the recombiners limiting the risk of hydrogen combustion.
Core melt with containment bypass	Passive provision and active provision for enhanced isolation as well and enhanced operational rules after identification of all bypass sources and verification.
Fuel melting within the SFP in the spent fuel building	Verification of all the design provision to control the cooling and water inventory of the SFP ensuring that there is always water covering and cooling the fuel assemblies.

II-1.2.7. Probabilistic assessment

The FA3 EPR deterministic approach is complemented by a comprehensive probabilistic approach made of a level 1 PSA and a level 2 PSA.

In addition to assessing the risk of fuel damage (core melt in the reactor or fuel melt, i.e. uncovered fuel in the SFP) the level 1 PSA confirms the absence of frequent sequences leading to fuel melt, which might occur when a series of lines of defence is lost by the failure of a single equipment. Therefore, this comforts the adequacy of the equipment reliability and the independence of equipment used in the first three levels of DiD.

The probability for core melt including all internal events and internal and external hazards is thus well below the 10^{-5}/r.y target.

The level 2 PSA assesses the risk of large and early releases to the environment as extremely unlikely, hence comforting a sufficient independence of the fourth level of DiD in comparison to the first three levels of DiD.

The European Utility Requirements [II-5] set a probabilistic quantitative design target value of 10^{-6}/r.y for sequences potentially leading to an early failure of the primary containment or leading to large releases; this target value is largely met for the EPR.

II-1.3. Design of FA3 EPR specific plant systems

Figure II-1 presents a schematic overview of the buildings of the FA3 EPR.

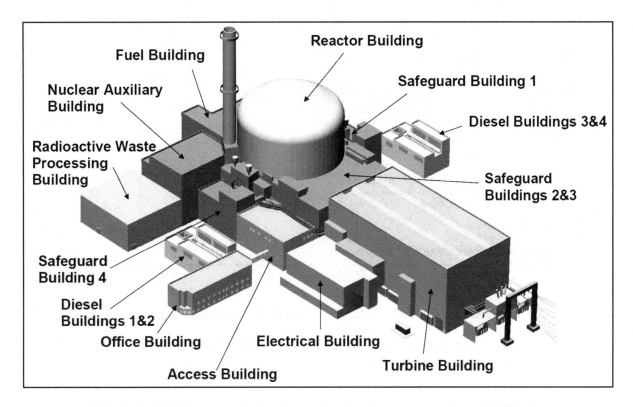

FIG. II-1. EPR 3D general view (reproduced with permission from Ref. [II-4]).

II-1.3.1. FA3 EPR safety systems

For a description of the safety systems provided to limit the consequences of design basis conditions DBC-2, DBC-3 and DBC-4, see Ref. [II-4].

II-1.3.2. Inherent safety features

The reliance on passive means for performing safety functions is not new. All existing PWRs include some passive features like accumulators, gravity-driven control rod insertion or natural circulation in the primary circuit. Besides these, additional passive features have been included in the EPR design, as described in the respective entry in IAEA's ARIS database [II-4].

II-1.3.3. FA3 EPR safety features to cope with severe accidents

The design target of the EPR is to limit the need for off-site emergency response actions (such as evacuation or relocation of the population) to the nearby plant vicinity, including in the hypothetical situation of a severe accident, where maintaining the integrity of the containment is essential. The EPR thus includes both preventive measures and mitigating features to prevent basemat melt-through and long term containment pressurization, to limit hydrogen deflagration and radioactive releases to the environment.

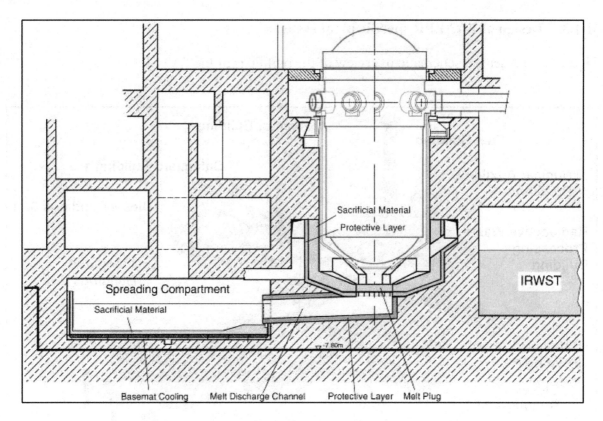

FIG. II-2. EPR core melt retention system (reproduced with permission from Ref. [II-4]). IRWST: in-containment refuelling water storage tank.

MCCI is prevented by spreading the corium in a spreading compartment provided with a protective layer and a special cooling device (FIG. II-2).

For more information on the severe accident provisions, see Ref. [II-4].

II-1.4. Consideration of external hazards

II-1.4.1. Identification and screening of external hazards for the design of nuclear power plants

The overall FA3 EPR approach to internal and external hazards can be summarized as:

- Hazard identification, with consideration of credible combinations, derived from the site evaluation process;
- Hazard impact quantification (e.g. specific loads and environmental conditions), and determination of the design basis (load case) of SSCs;
- Design verification against hazards to confirm that the safety objectives are achieved, practically performed building by building, specifically to each hazard characteristics, with the use of deterministic studies, such as building and equipment response, functional impact analysis including consideration of consequential internal events (for instance, identification of internal events induced by an initiating internal fire hazard).

Basically, the hazard approach objective is to demonstrate that despite the hazard event, the fundamental safety functions would be ensured and that the functions required to bring the reactor to a safe state can be carried out satisfactorily, so that radiological releases would be limited and remain acceptable. Practically, the protection is achieved by appropriate sizing, redundancy, diversity and segregation, through layout rules and supported by the application of relevant codes and standards.

II-1.4.2. Consideration of levels of natural hazards exceeding those considered for the design

The external hazard level or load quantification accounts for all foreseeable phenomena in direct relation to the hazard or consequential to it (e.g. external flooding is accounting for maximal tide level, waves, wind, rain), aiming at defining a bounding case where cliff edge effects are considered (e.g. an increase of the sea level does not compromise the external flooding protection). This bounding case has therefore intrinsic margins. On this basis, the design accounts for additional margins:

- Codes and standards used to design the SSCs introduce provisional coefficients as part of the state of the art and the potential for cliff edge effects with respect to the load cases, including, but not limited to, the following:
    - The French Association for Nuclear Steam Supply Systems Equipment Construction Rules (AFCEN);
    - The design and construction rules for the mechanical components of PWR nuclear islands (RCC-M);
    - The design and construction rules for electrical and I&C systems and equipment (RCC-E).
- Provision via decoupling assumptions defined early in the design development process (while the final values might not yet be known). This generally ends up with the initial assumption providing margin in comparison to the final required value;
- Combination of loads (e.g. seismic, thermal-hydraulic transients) to design the SSCs are considered in a conservative manner.

However, in addition to such approaches, as part of the lessons learned from the Fukushima Daiichi accident, the FA3 EPR has been reinforced against extreme natural hazards exceeding the hazard design basis.

The approach is to define a set of design provisions and organizational means, aiming at preventing large radioactive releases and durable effects on the environment if such extreme events were to occur, and to design them against these extreme events, or at least to demonstrate their availability in such conditions. This so-called 'hardened safety core approach' considers the following hazards and their relevant combinations:

- Extreme earthquake;
- Extreme external flooding (including heavy rainfalls) and any natural phenomenon concurring to or consequential to (such as extreme winds, lightning, and hail);
- Tornado.

The set of SSCs of the hardened safety core have the following main functions:

- Preventing core melt by providing appropriate means of cooling the reactor core and remove decay heat outside of the containment, if possible, to pressurize the primary circuit (i.e. cooling the reactor via the secondary circuit is possible and, therefore, the integrity of the reactor coolant primary boundary is maintained);
- Ensuring that there is no uncovered fuel in the SFP;

- Maintaining the integrity of the third barrier (containment building) when initially not affected (equipment hatch and airlocks closed) and prevent bypasses of this barrier;
- Limiting radioactive releases to the environment, even in case of a severe accident and a failure of the reactor vessel (core melt-through).

In addition, design provisions are integrated in the FA3 EPR design for the use of non-permanent or mobile equipment. This includes the provision to inject cold water inside the containment and remove heat in case of a severe accident.

II-2. GUIDE N° 22 ON THE DESIGN OF PRESSURIZED WATER REACTORS

The guide [II-3] was developed jointly by ASN and the Institut de Radioprotection et de Sûreté Nucléaire (IRSN) and presents recommendations for the design of PWRs. It specifically considers the safety requirements for reactor design taken from the IAEA publications and the reference levels, safety objectives and recommendations published by WENRA [II-6].

II-2.1. Purpose of the guide

The guide [II-3] presents the recommendations of ASN and IRSN for the design of PWRs. Its primary objective is to address the prevention of radiological incidents and accidents and the limitation of their consequences. It also addresses other aspects associated with the management of non-radiological risks or the adverse effects that will result from the facility operation.

The guide was drawn up based on knowledge resulting from examinations performed on the NPPs in operation, under construction, or at the project stage in France. The guide considers the lessons drawn from the reviews of the technical files submitted to ASN by the French applicants which have highlighted the relevance of certain practices. It will be updated regularly to incorporate new knowledge, experience feedback (as much from its application to concrete examples as from operation of the facilities), recommendations made by international organizations and new practices. To take such changes into account, ASN may issue additional — or even alternative — recommendations before the next revision of the guide.

The guide is intended for future licensees of PWRs in France, responsible for controlling the risks and inconveniences that the installation can present in accordance with article L. 593-6 of the Environment Code of France, as well as for the authors of specifications and the designers of such installations without prejudice to the responsibilities of the nuclear pressure equipment manufacturers provided for by the regulations.

The guide sets out, through inset text boxes, the regulatory requirements to consider in the design, while the main body of the text presents the recommendations that enable these requirements to be satisfied, as much from the technical aspects as from the relevant organizational and human factors aiming to protect the interests mentioned in the first paragraph of article L. 593-1 of the Environment Code.

II-2.2. Scope of the guide

The guide applies to PWRs. It essentially addresses the prevention of radiological incidents and accidents and the limitation of their consequences, knowing that other aspects (relating to the management of non-radiological risks or the adverse drawbacks that will result from normal operation of the installation, radiation protection and safety of workers) are to be considered in the design of PWRs.

As the guide applies primarily to the design of new-generation PWRs, its recommendations may also be used, for reference, to seek improvements to be made to reactors in operation, for example during their periodic safety reviews, in accordance with article L. 593-18 of the Environment Code and articles 8a and 8c introduced by the Council Directive 2014/87/EURATOM of 8th July 2014 [II-7].

II-2.3. Status of the guide

At the date of its publication, the guide shall be considered in priority for the PWRs whose creation authorization decree has not yet been issued.

Compliance with the recommendations of the guide is a satisfactory way of meeting the regulatory requirements concerning nuclear safety. It is nevertheless possible to depart from the recommendation if it is proved that the regulatory requirements are satisfied by other means. If there are no recommendations on a specific subject, the acceptability of the licensee's proposal for a given project will be assessed in the examination of the file concerning that project.

The guide underwent a consultation by the stakeholders, including the basic nuclear installation licensees, in September 2016, followed by an examination by the French advisory committee of experts for nuclear reactors (GPR) with the participation of members of the French advisory committee of experts for nuclear pressure equipment (GPESPN).

REFERENCES TO ANNEX II

[II-1]   AUTORITE DE SÛRETÉ NUCLÉAIRE, Technical guidelines for the design and construction of the next generation of Nuclear Power Plants with Pressurized Water Reactors. Adopted during the GPR/German experts plenary meetings held on October 19th and 26th (2000).

[II-2]   INTERNATIONAL ATOMIC ENERGY AGENCY, Safety of Nuclear Power Plants: Design, IAEA Safety Standards Series No. SSR-2/1 (Rev. 1), IAEA, Vienna (2016).

[II-3]   AUTORITE DE SÛRETÉ NUCLÉAIRE, Guide de l'ASN n° 22: Conception des réacteurs à eau sous pression (2017).

[II-4]   INTERNATIONAL ATOMIC ENERGY AGENCY, Advanced Reactors Information System (ARIS) Database, The Evolutionary Power Reactor (2011), https://aris.iaea.org/PDF/EPR.pdf.

[II-5]   EUROPEAN UTILITY REQUIREMENTS ORGANISATION, European Utility Requirements for LWR Nuclear Power Plants, Revision E (2017).

[II-6]   WESTERN EUROPEAN NUCLEAR REGULATORS' ASSOCIATION, Safety of New NPP Designs, WENRA Reactor Harmonisation Working Group (2013).

[II-7]   EUROPEAN COMMISSION, Council Directive 2014/87/EURATOM of 8 July 2014 amending Directive 2009/71/Euratom establishing a Community framework for the nuclear safety of nuclear installations, Official Journal of the European Union, L 219, 25 July 2014.

Annex III

JAPAN

III-1. INTRODUCTION

This annex provides an overview of the new regulations, covering an extended consideration of external events in the design of NPPs, the strategy of DiD, enhancement of measures against DECs and beyond DECs, the use of NPE for severe accident management, measures to address large damage of NPPs, and measures to suppress dispersion of radioactive material. More detailed information is provided in the national report of Japan for the 8th Review Meeting of the Contracting Parties to the Convention on Nuclear Safety [III-1].

III-2. OVERVIEW OF THE NEW REGULATIONS

The Nuclear Regulation Authority (NRA) updated the regulatory requirements for NPP design and operation in July 2013 [III-2], incorporating the lessons learned from TEPCO's Fukushima Daiichi Nuclear Power Station accident.

The regulatory requirements are based on the concepts of DiD, which provide multiple (or multilayered) independent and effective countermeasures to PIEs. The requirements reinforce the provisions and the countermeasures against natural phenomena and other events (such as fire) that trigger CCF. Moreover, they require measures to prevent core damages and containment vessel failures in case of a severe accident, measures for suppression of dispersion of radioactive material, and protective measures against intentional aircraft crashes.

The basic policies for designing measures against severe accidents and malevolent acts are as follows:

- Protective measures result from multiple stages such as 'prevention of core damage', 'securing the containment function', and 'suppression of dispersion of radioactive material';
- Further enhancement of reliability results from the combination of permanent and NPE, while primarily based on the use of NPE;
- Reinforcement of cooling measures in the spent fuel storage pool;
- Enhancement of the emergency response centre durability, reliability and durability of communication systems, and reliability and persistence of measurement systems including those in spent fuel storage pool (reinforcement of command communications and measurement systems);
- Preparation of procedures, securement of personnel, and implementation of trainings are required since it is important for hardware (facilities) and software (on-site work) to function integrally;
- As countermeasures against intentional airplane crashes, distributed storage and preparation of connection of NPE are required. The SSF is introduced as a backup measure for reliability enhancement.

Table III-1 shows the update of national requirements in Japan after TEPCO's Fukushima Daiichi accident.

TABLE III-1. UPDATE OF REGULATORY REQUIREMENTS IN JAPAN AFTER TEPCO'S FUKUSHIMA-DAIICHI ACCIDENT (REPRODUCED WITH PERMISSION FROM REF. [III-5])

Suppression of radioactive materials dispersal	Beyond DEC	*. Regulatory requirements were defined only in the Design Basis before 1F Accident.
Specialized Safety Facility		
Prevention of CV failure	DECs	
Prevention of core damage		
Natural phenomena	The 2nd and 3rd layers of DiD	Reinforced.
Fire, Internal flooding		
Reliability		
Reliability of power supply		
Ultimate heat sink		
Function of other SCCs		
Seismic/Tsunami resistance		

III-3. BROADENED CONSIDERATION OF EXTERNAL EVENTS (E.G. EARTHQUAKE, TSUNAMI AND OTHER EVENTS)

III-3.1. Regulatory requirements for external events

Under the NRA Ordinance on Standards for Installation Permit [III-3], the regulatory requirements for external events (both natural and human induced events) were significantly reinforced in response to the lessons learned from the TEPCO's Fukushima Daiichi accident, as summarized below:

- Buildings and SSCs important in terms of seismic design are required to be installed on the ground without an outcrop potentially subject to a fault, because buildings and internal equipment may be damaged not only due to motion but also due to ground displacement or deformation. Additionally, the standards for determining capable faults (faults likely to becoming capable in the future) are defined as follows: The faults are identified as capable if it is not possible to deny fault activities after the late Pleistocene (about 120 000 to 130 000 years ago). If necessary, evaluation of fault activities has to be made by going back to the middle Pleistocene (about 400 000 years ago).
- For prevention of damage caused by earthquakes, it is required that safety functions of the buildings and SSCs with importance in terms of seismic design are not lost against the seismic force and potential slope collapses generated by the design basis ground motion. The design basis ground motion is defined considering the latest scientific and technical knowledge from a seismological and earthquake engineering standpoint (such as geology, geological structure, soil structure, earthquake activities on and around the site), and it is required to define both:
  i. Seismic ground motions with identifying seismic sources in each site;
  ii. Seismic ground motions without identifying seismic sources.
- For the first type, the definition has to be made by selecting several earthquake types (such as continental crust earthquake, interplate earthquake, and oceanic interpolate earthquake), that are expected to have large influence on the site, considering the uncertainties and reflecting the propagation characteristics of earthquake waves. For the second type, the identification has to be made by collecting the observation records from

the past earthquakes that occurred in the continental crust with seismic sources related to capable faults and by considering the ground characteristics of the site. In terms of propagation characteristics of earthquake waves, it is required to evaluate the three-dimensional subsurface structure under a site, considering the possibility that a seismic ground motion is amplified due to the subsurface structure under the site. It is required to assess the exceedance probability from the design basis ground motion.

- For prevention of damage caused by tsunamis, it is required to identify as design basis the tsunami characterized by a level exceeding the past maximum value and to install protective facilities against the design basis tsunami such as protective seawalls to prevent water inundation into the site or tide gates to prevent water inundation into buildings. Tsunami protective facilities has to be of the S class, the highest class in the seismic design classification, so that the flooding prevention functions are not lost due to earthquakes. For the design basis tsunami, the identification should be made considering the latest scientific and technical knowledge of tsunamis that should be postulated from a seismological perspective (such as ocean floor topography, geological structure, seismic activities). As for the mechanisms that may cause tsunamis, in addition to the earthquakes (interplate earthquakes, oceanic interpolate earthquakes, and continental-crustal earthquakes due to capable faults in a relevant sea area), landslide, slope collapse, and other mechanisms and a combination of these should be selected. Furthermore, due consideration to uncertainties should be made in the numerical analyses. The exceedance probability for the design basis tsunami should be assessed.

- Safety functions should not be lost in case of occurrence of natural phenomena other than earthquakes and tsunamis (such as floods, typhoons, tornados, freeze, precipitation, accumulation of snow, lightning, landslides, influences of volcanoes, biological events or forest fires, or a combination of these). For human induced events (except intentional ones), it is also required not to lose the safety functions in case of occurrence of missiles (e.g. airplane crash), collapse of dams, explosions, fires in nearby factories, toxic gas, collision of ships, electromagnetic interferences, based on situations on and around a given site.

- A SSF should be built to avoid releases of radioactive materials in the event of an intentional large airplane crash or malevolent acts. The SSF is a facility that can be used until external support becomes available in the event of, for example, an intentional airplane crash, and which has the necessary equipment for preventing damages to the containment vessel; the SSF has not to lose its function even in the event of an airplane crash into the reactor building. Moreover, robustness has to be enhanced against motions exceeding the design basis ground motion to a certain degree.

III-3.2. Regulatory guides

The NRA has developed several regulatory guides on the topic, including:

- A guide for the review of design basis earthquake and seismic resistance design;
- A guide for the review of design basis tsunami and tsunami resistance design;
- A guide for the review of foundation grounds and the assessment of slope stability;
- A guide for the assessment of volcanic hazards;
- A guide for the assessment of tornado hazards;
- A guide for the assessment of external fires.

These regulatory guides are posted on the NRA web site in Japanese.

III-4. DEFENCE IN DEPTH STRATEGY

III-4.1. Basic policy on defence in depth in Japan

In the past, before the NRA's regulatory requirements were updated, the DiD concept was stated in the Reactor Regulation Act and Regulatory Guides issued by the Nuclear Safety Commission [III-4], and described as follows;

- For the first layer, the objective is to ensure high reliability commensurate with the importance of SSCs to prevent occurrence of abnormality.
- For the second layer, the objective is to take necessary measures for early finding of abnormality and shut down the reactor to prevent escalation of abnormality.
- For the third layer, the objective in case of DBAs is to avoid a severe damage to the core and to maintain a coolable geometry and to avoid radiological consequences to the public in the vicinity of the NPP.

In the new regulatory requirements issued by the NRA [III-3], measures to eliminate CCF are significantly strengthened, based on the lessons learned from the TEPCO's Fukushima Daiichi accident. In addition to the abovementioned requirements, measures for preventing severe core damage are required in the case of loss of safety functions mitigating DBAs, and measures for preventing containment vessel failure are also required even in case of severe core damage. Furthermore, mitigating measures assuming containment vessel failure are required. In accordance with the updated requirements, measures for preventing large damage of NPPs due to extreme natural hazards, intentional airplane crashes or other malevolent acts have also to be implemented.

According to [III-3] each layer of DiD has to perform its function effectively and independently from the other levels.

III-4.2. Requirements for each layer of defence in depth

III-4.2.1. Prevention of abnormality

To prevent abnormality, it is required to ensure high reliability, to have sufficient safety margin, to have intrinsic feedback characteristics in reactivity control, and to prevent operator error. Fail-safe design and interlock function are used to deal with operator error or a failure.

In the new regulatory requirements [III-3], measures for seismic design, prevention against tsunami, reliability of power supply and fire protection are strengthened, and measures for prevention against internal flooding, volcano eruption, tornado and forest fire are newly required. Regulatory requirements for external events are reported in Section III-3 of this annex.

III-4.2.2. Prevention of escalation from abnormality

Measures to prevent anticipated transients from escalating to an accident are required, such as the design of specific systems and mechanisms, and establishing operational procedure to return to a safe state.

III-4.2.3. Mitigation of design basis accidents

In case of escalation of anticipated transients or PIEs to DBAs, it is required that the core is not severely damaged and is able to maintain a coolable geometry, avoiding radiological releases to the public in the vicinity of the NPP by intrinsic safety and ESFs.

III-4.2.4.   Prevention of severe core damage in design extension conditions without significant fuel degradation

Licensees are required to confirm the effectiveness of measures to prevent severe core damage in DECs without significant fuel degradation.

DECs without significant fuel degradation are identified as 'postulated accident sequence groups'. The NRA Reactor Establishment Permit Ordinance [III-3], taking benefit of results from R&D, identifies accident sequence groups which cover most of the accident sequences with potential significant core damage as 'designated accident sequence groups' as shown in the Table III-2.

TABLE III-2. DESIGNATED ACCIDENT SEQUENCE GROUPS (PREVENTION OF SEVERE CORE DAMAGE)

BWR	PWR
Loss of high-pressure and low-pressure water injection function	Loss of heat removal function of secondary cooling system
Loss of high-pressure water injection and depressurization function	Loss of AC power
Loss of all AC power	Loss of component cooling function
Loss of decay heat removal function	Loss of containment vessel heat removal function
Loss of reactor shutdown function	Loss of reactor shutdown function
Loss of water injection during LOCA	Loss of ECCS water injection function
Containment vessel bypass (Interface system LOCA)	Loss of ECCS recirculation function
	Containment vessel bypass (Interface system LOCA, SGTR)

Considering the difference of each plant, internal events are evaluated by applying probabilistic risk assessment (PRA), and external events are evaluated by PRA or other applicable means. As a result, accident sequence groups with significant frequency or impact are added into a 'postulated accident sequence group' (if not already included in the 'designated accident sequence group').

In the next step, important accident sequences are identified in each of the postulated accident sequence group from the perspective of the number of SSCs that lose their function simultaneously, the time margin, the SSC capacity necessary to prevent core damage, and based on the representativity of the characteristic of the specific accident sequence group. An evaluation of the effectiveness using numerical simulation modelling analysis (e.g. maximum temperature of fuel cladding is below 1200°C) is used to confirm that the relevant SSCs designed for prevention of severe core damage in such accident sequences meet the requirement. Operational measures (e.g. necessary personnel and oil fuel, if any) are also assessed. The equipment required to address such DECs has to meet the following regulatory requirements:

- The equipment has not to lose its function due to a CCF simultaneously to safety systems designed for DBAs.
- The equipment has to be furnished with anti-seismic function.

In addition to these requirements, high reliability is required for permanently installed equipment.

III-4.2.5.   Prevention of containment vessel failure in design extension conditions with core melting

Licensees are required to confirm the effectiveness of measures to prevent containment vessel failure in the case of DECs with core melting.

DECs with core melting are identified as 'containment vessel failure modes'. The NRA Reactor Establishment Permit Ordinance [III-3], taking benefit of results from R&D, stipulates 'designated containment vessel failure mode' as the typical containment vessel failure modes. Static loads are typically due to internal atmospheric pressure or temperature (damage by overpressurization or overheating of the containment vessel) caused by the following events:

- HPME/direct heating of the containment vessel atmosphere;
- Ex-vessel FCI;
- Hydrogen explosion;
- Direct contact with containment vessel (shell attack);
- Molten core concrete interactions (MCCI).

Considering the difference of each plant, internal events are evaluated by applying PRA and external events are evaluated by PRA or other applicable means to identify a containment vessel failure mode based on the plant characteristics. As a result, containment vessel failure modes with significant frequency of occurrence or impact are added into a 'postulated containment vessel failure mode' (if not already included in the 'designated containment vessel failure mode').

In the first step, for every postulated containment vessel failure mode, a severe accident sequence is identified as an evaluated accident sequence among containment vessel failure sequences based on the results of PRA. Subsequently, an evaluation of the effectiveness is performed to confirm that the equipment designed for coping with severe accident can prevent containment vessel damage in such accident sequences by meeting the evaluation requirements (e.g. operating pressure) obtained through analysis by numerical simulation codes. Operational measures (e.g. necessary personnel and oil fuel, if any) are also assessed.

The Ref. [III-3] for evaluating the effectiveness requires confirmation that the pressure and temperature values fulfil the acceptance criteria, that radioactive releases are as low as possible, and that the release of Cs-137 is lower than 100 TBq.

Equipment required to mitigate DECs with core melting has tomeet the following regulatory requirements:

- The equipment has to perform its function in accident conditions.

- Redundancy or diversity, independence and physical separation in different locations have tobe ensured if the equipment to address DBAs has no similar function (e.g. water injection to containment vessel bottom, prevention of hydrogen explosion).
- Equipment has to be seismically qualified.

In addition to these requirements, high reliability is required to permanently installed equipment. For NPE, meeting general industrial standards and multiple deployment of equipment (e.g. water injection, power source) is required.

III-4.2.6.   Measures to suppress dispersion of radioactive material

The NRA Ordinance on Standards for Installation Permit [III-3] requires measures for prevention of severe core damage and containment vessel failures. The NRA Ordinance requires equipment to suppress the dispersion of radioactive material off-site, based on appropriate analysis of the dispersion mode, to prevent abnormal levels of release of radioactive material into the environment, even if assuming severe core damage and containment vessel failure in the late phases of a DEC. For example, a water cannon is required to suppress dispersion of radioactive material in aerosol form leaking from the reactor building.

III-4.2.7.   Measures to address large scale damage of nuclear power plants

Large scale damage of nuclear facilities is the level of destruction caused by extreme natural hazards (i.e. natural hazards beyond the design basis), intentional airplane crashes, or other malevolent acts.

In the NRA Ordinance [III-3], the use of NPE and a SSF are required in order to prevent such large scale damage:

a) Use of NPE: airplane crash leads to severe destruction of certain areas of NPPs, causing large scale damage. In this case, it is important to foresee measures not based on a specific accident sequence but to avoid losing all mitigative measures aiming at decreasing radioactive releases, should such a scenario occur.
   In case of natural hazards of extreme magnitude (beyond design basis) or a large airplane crash, it is required to avoid the simultaneous unavailability of NPE, through the diversification of such equipment. In practical terms, this might imply:

   - The access route (such as roads) has to be repaired by heavy machinery stored in diversified locations if the access route is destroyed by natural hazards beyond the design basis;
   - Ensuring the existence and availability of connecting points in the opposite side of the site, to be able to connect NPE (e.g. feedwater pump or power source) at least on a side, in the case connection points are lost due to an airplane crash into the other side of the reactor building.

b) SSF: this "shall be equipped with adequate measures for preventing the loss of necessary function due to the intentional crashing of a large airplane into the reactor building and other malevolent acts" [III-3]. Practical requirements are:
   - To ensure sufficient distance (e.g. more than 100 m between the SSF and the reactor building) to prevent simultaneous failure of both facilities;
   - The SSF "shall be equipped with a robust structure that can withstand an intentional airplane crash" [III-3].

Licensees have to demonstrate that the equipment will fulfil its function by performing a structural evaluation of the building response as well as a functional evaluation of the equipment in case of airplane crash (necessary inputs are the airplane characteristics and the exact crash location).

'Equipment to prevent containment vessel failure' has to be equipped in the SSF. Practical requirements are:

- To ensure depressurization function for reactor coolant pressure boundaries (e.g. equipment for reactor depressurization operation from the emergency control room);
- Cooling function of the molten core in the reactor (e.g. equipment for injecting low pressure water into the reactor);
- Function for cooling the molten core which might have fallen outside the bottom of the containment vessel (e.g. equipment for water injection into the bottom of the containment vessel);
- Containment vessel functions of cooling, depressurization, and radioactive material reduction (e.g. equipment for injecting water into containment vessel sprays);
- Containment vessel functions of heat removal and depressurization (e.g. filtered venting);
- Function for preventing containment vessel failure by hydrogen explosion (e.g. equipment for control of hydrogen concentration);
- Support functions (e.g. equipment for power source, instrumentation, and communication);
- An emergency control room function to control the above mentioned functions.

III-5. CONCLUSION

The new regulation significantly enhances the safety of NPPs by expanding the consideration of external events, enhancement of measures against DECs and beyond DECs, use of NPE for severe accident management, measures to address large damage of NPPs, and measures to suppress dispersion of radioactive material off-site.

REFERENCES TO ANNEX III

[III-1] NUCLEAR REGULATION AUTHORITY, Convention on Nuclear Safety – National Report of Japan for 8th Review Meeting, NRA, Tokyo, Japan (2019), https://www.nsr.go.jp/data/000280849.pdf.

[III-2] NUCLEAR REGULATION AUTHORITY, New Regulatory Requirements (2013), https://www.nsr.go.jp/english/regulatory/index.html.

[III-3] NUCLEAR REGULATION AUTHORITY, Ordinance Prescribing Standards for the Location, Structures, and Equipment of Commercial Power Reactors and their Auxiliary Facilities, Reactor Establishment Permit Ordinance (2013), https://www.nsr.go.jp/english/regulatory/index.html.

[III-4] Act on the Regulation of Nuclear Source Material, Nuclear Fuel Material and Reactors, No. 166, 10 June 1957, as amended (Reactor Regulation Act).

[III-5] INTERNATIONAL ATOMIC ENERGY AGENCY, Experiences in Implementing Safety Improvements at Existing Nuclear Power Plants, IAEA-TECDOC-1894, IAEA, Vienna (2020).

Annex IV

REPUBLIC OF KOREA, APR1000

IV-1. GENERAL INTRODUCTION OF THE APR1000

The APR1000 is a 1000 MW(e) Generation III+ pressurized water reactor designed to comply with the up-to-date international safety requirements of IAEA, WENRA and U.S. NRC. It was developed based on the technology of Optimized Power Reactor 1000 MW (OPR1000) and Advanced Power Reactor 1400 MW (APR1400) that have been proven by successful construction, operation and design certification in the Republic of Korea, United Arab Emirates and United States. It also incorporates the advanced design features of the Advanced Power Reactor Plus (APR+) and the European-Advanced Power Reactor (EU-APR), which designs have been licensed by the Korean regulatory authority and received certification of compliance from the European Utility Requirements Organization.

This annex provides the design features of the APR1000 with respect to the safety improvements aligned with the latest requirements established in IAEA Safety Standards Series No. SSR-2/1 (Rev. 1) [IV-1] and the associated technical guidance such as WENRA [IV-2] and European Utility Requirements Rev. E [IV-3].

IV-2. INDEPENDENCE OF LEVELS OF DEFENCE IN DEPTH

IV-2.1. Implementation of defence in depth principles

The DiD principle is implemented within the design of the APR1000 by adapting five levels of protection and with multiple physical barriers preventing the release of radioactive materials to the environment. The concept of DiD focuses primarily on preventing deviation from normal operation and, if this is not successful, in mitigating the potential consequences and, thus, preventing the plant conditions from escalating to more serious events by providing independent design provisions dedicated for each plant state.

The APR1000 design divides the plant states in five different categories, based on the frequency of occurrence, as follows:

1. Normal operation: DiD level 1;
2. Anticipated operational occurrences (AOOs): DiD level 2 ($f > 10^{-2}$/r.y);
3. Design basis accidents (DBAs): DiD level 3a:
   i. DBA-1: $10^{-4} < f < 10^{-2}$/r.y;
   ii. DBA-2: $10^{-6} < f < 10^{-4}$/r.y;
4. Design extension conditions (DECs):
   i. DECs without significant fuel degradation (also called DEC-A): multiple failures: DiD level 3b ($10^{-6} < f < 10^{-4}$/r.y);
   ii. DECs with core melting (also called DEC-B): severe accidents: DiD level 4 ($f < 10^{-6}$/r.y).

Systems used for each level are designed, to the extent practicable, to be independent from those used for other levels of DiD.

Two kinds of safety targets, i.e. deterministic and probabilistic, are established in the design of the APR1000:

a) The qualitative deterministic safety targets are set to meet the safety principle of IAEA Safety Standards Series No. SF-1, Fundamental Safety Principles [IV-4]:
   - For normal operation and AOOs, the targets are set such that the radiological impact to the public is 'negligible';
   - For DBAs and DECs without core melt (DEC-A or multiple failure), the targets are set to have 'no or minor radiological impact';
   - For DEC-B, which means severe accident conditions which occurrence is not 'practically eliminated', the targets are set to require 'only limited protective measures in terms of area and time';
   - For DEC-B conditions that should be 'practically eliminated', the radiological targets are not required, but separate acceptance criteria of either physical impossibility or extremely unlikeliness are applied;
b) Considering the current international nuclear power industry's technology, the CDF target for all the internal and external events is set to 10^{-5}/r.y, while 10^{-6}/r.y is the target for the LRF.

Figure IV-1 schematically presents the safety targets of the APR1000.

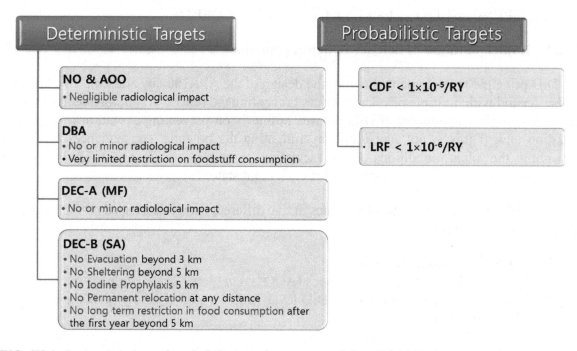

FIG. IV-1. Deterministic and probabilistic safety targets of the APR1000 (courtesy of Sangho Kang, KEPCO Engineering and Construction).

The APR1000 design implements the DiD concept by establishing safety objectives, acceptance criteria for three physical barriers and radiological criteria assigned for each DiD level.

In order to balance the risk for the wide frequency range of DBAs, these are further broken down into DBA-1 and DBA-2 according to their frequency of occurrence.

For DEC-A, the following types of events are included:

- AOO or most frequent DBA-1 events combined with postulated CCF of redundant trains of required safety systems;
- Complex or specific scenarios including CCF of systems needed to fulfil the fundamental safety functions in normal operation.

In the design of the APR1000, the safety systems used for a particular DiD level are designed independently from those used for different DiD levels, as far as reasonably practicable. While many of the safety systems are used only for one dedicated DiD level, there are several systems used in two or more DiD levels, based on the following justification of the European Utility Requirements Organization:

- Since the events considered in one DiD level do not always result from a failure of the systems in the previous level of defence, it is not feasible to design an NPP with independent systems for all PIEs at each level of DiD.
- There are several SSCs that are credited in more than one level of DiD and, thus, it is neither possible nor beneficial for safety to provide each SSC to all levels of DiD.
- The emergency AC power supply belonging to DiD level 3a may be used also in DiD level 2. An additional diverse emergency AC power supply will be designed for DiD level 3b because the CCF of the primary (non-diverse) emergency power supply is postulated. The alternate power source on DiD level 3b may be also used for DiD level 4. The rationale for this is that additional independent on-site provisions are not likely to significantly increase the reliability of the emergency AC power supply.

IV-2.2. Safety systems for anticipated operational occurrences and design basis accidents

The ESFs of the APR1000 are designed to provide safety functions for AOOs and DBAs, such as control of reactivity, heat removal from the core and confinement of radioactive materials. However, these systems can also be used in some of DEC-A events if they are not impaired by the corresponding DEC-A events. Main safety systems designed as ESFs are the following ones:

- Safety injection system (SIS);
- Shutdown cooling system / containment spray system (CSS);
- Passive AFS;
- Safety depressurization and vent system;
- In-containment refuelling water storage system;
- Containment isolation system (CIS);
- Component cooling water system (CCWS);
- Essential chilled water system (ECWS);
- Essential service water system (ESWS).

The supporting I&C systems, such as the reactor protection system, the ESF component control system and the auxiliary power system (which power is supplied by EDGs), are designed to ensure reliable operation of the above ESFs.

In addition, the safety classified HVAC systems — such as the auxiliary building emergency ventilation system, the fuel handling area emergency ventilation system, and the secondary containment emergency ventilation system — are provided to confine the radioactive materials.

The active safety systems used to reach and maintain a controlled and/or a safe state after an AOO or a DBA are designed to have four train redundancy considering the SFC and on-line maintenance, which means (N+2) redundancy. Each train and its components are located in four physically separated quadrants inside the auxiliary building to secure their safety functions in presence of internal and external hazards.

The APR1000 design provides the safety depressurization and vent system, which enables a feed and bleed operation and depressurizes the RCS rapidly by venting the pressurizer through the pilot operated SRVs.

The in-containment refuelling water storage tank is located inside the containment, so that the injected emergency cooling water returns to that storage tank. It provides several safety functions such as water source for refuelling as well as for safety injection, containment spray, and feed and bleed operations. It is also used as a heat sink for discharged steam from the ERDS for prevention of HPME and as a source of water supply to PECS, which is designed to prevent MCCI in severe accident conditions.

The shutdown cooling system is combined with the CSS, so that both functions of residual heat removal and containment cooling can be fulfilled with (N+2) redundancy.

The passive AFS provides core heat removal function (even in the absence of power supply) by extracting the steam from the SGs through the passive condensation heat exchangers and returning the condensate to the SG by natural convection.

The CIS provides the means of isolating fluid systems that pass through the containment penetrations in order to avoid radioactive releases through containment bypass following an accident.

IV-2.3. Safety features for design extension conditions without significant fuel degradation (DEC-A)

The diverse safety features of the APR1000 provide fundamental safety functions in case of DEC-A.

The emergency boration system (EBS) provides a diverse mean of reactor trip by injecting high concentration borated water into the RCS following a ATWS caused by mechanical failure of control rods.

The diverse CSS, which is primarily designed for containment cooling in DEC-B conditions, can be also used as RHR system in case of a loss of RHR with its supporting safety features, namely the diverse CCWS and the diverse essential service water system.

The diverse SFP cooling system provides a function to remove the decay heat generated by the spent fuels stored in the SFP in the event of a loss of SFP cooling.

Associated I&C systems, such as the diverse protection system and the diverse component control system, and the diverse HVAC system, such as the diverse essential chilled water system, which controls and cools down the above diverse safety features, are also independent from those designed for DBAs. The alternate AC DG, which is a diverse power source from the EDGs, supplies the electrical power to essential components to cope with SBO condition.

IV-2.4. Safety features for design extension conditions with core melting (DEC-B)

The APR1000 provides severe accident mitigation safety features that are independent from those used for other DiD levels, aiming at limiting the consequences of DECs with core melting:

- ERDS;
- PECS;
- Diverse CSS;
- HMS.

The associated I&C system (alternate monitoring and control system) and the electrical power from the alternate AC DG are also independently provided.

The ERDS is independent from the safety depressurization and vent system and rapidly depressurizes the RCS to eliminate the occurrence of HPME under DEC-B conditions.

The PECS (also called 'core catcher') is located in the reactor cavity to capture and cool-down the discharged molten core debris and, thus, protects the cooling structures and the containment basemat.

The diverse CSS decreases the containment pressure and temperature in DEC-B conditions by condensing the steam generated in the containment and reduces the potential for further pressure increase by removing decay heat from the containment atmosphere and from core debris in the reactor cavity. The diverse CSS also has the capability to remove the fission products in the containment atmosphere in severe accident conditions. The diverse CSS also provides the means to supply cooling water to PECS for long term recirculation cooling of corium.

The HMS is designed to control combustible gases inside the containment within the acceptable limits with PARs which reduce hydrogen generation during DEC-B conditions.

IV-3. 'PRACTICAL ELIMINATION'

The concept of 'practical elimination' of early or large radioactive releases is incorporated in the design of the APR1000. The 'practical elimination' implementation is fulfilled by preventing such conditions followed by a demonstration. Three steps applied to the APR1000 to achieve this goal are as follows:

- Identify phenomena that should be 'practically eliminated';
- Provide design provisions to prevent the occurrence of each phenomenon;
- Demonstrate 'practical elimination' by either physical impossibility or extremely unlikeliness with a high level of confidence.

The phenomena that are to be 'practically eliminated' are identified through the PSA approach and confirmed by comparison with the international guidance provided within the IAEA (see Section 4) and above-mentioned WENRA and European Utility Requirements references. The APR1000 design identifies ten phenomena to be 'practically eliminated' and provides appropriate design provisions to prevent the occurrence of each phenomenon. Approaches to the demonstration of 'practical elimination' of the APR1000 are briefly summarized in Fig. IV-2 and discussed in detail in the Appendix.

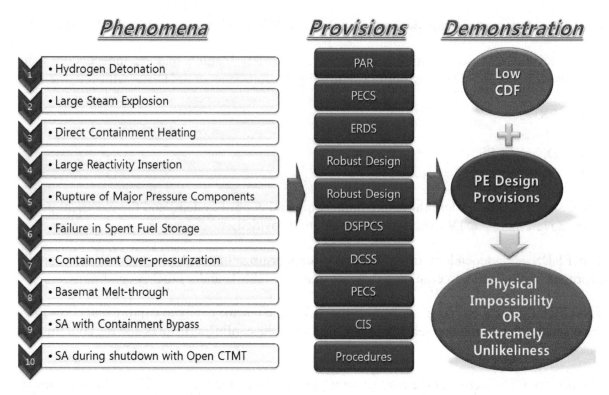

FIG. IV-2. Concept of demonstration of 'practical elimination' of the APR1000 (courtesy of Sangho Kang, KEPCO Engineering and Construction).

As a quantitative target for the demonstration of 'extremely unlikeliness', the APR1000 standard design sets a probabilistic target value of 10^{-7}/r.y for each sequence object of a demonstration of 'practical elimination'. It is derived from the level 3 PSA for Korean NPPs, where the large or early release frequency of 10^{-7}/r.y from level 2 PSA is equivalent to a fatality of 5.0×10^{-9}/r.y in level 3 PSA. The fatality value of 5.0×10^{-9}/y. corresponds to a hundredth of the trivial risk level of 5.0×10^{-7}/y. This means that, if the frequency of a phenomenon that may cause an early or large release is limited below 1.0×10^{-7}/r.y, its risk of fatality is less than 5.0×10^{-9}/year, hence a negligeable risk.

IV-4. DESIGN FOR EXTERNAL HAZARDS

The concept of RSEH has been introduced in the design of the APR1000 to address the natural external hazards exceeding the design basis external hazard (DBEH). The key requirement is that the plant should have adequate safety margin to avoid cliff edge effects with respect to early or large radioactive releases even in the event of RSEH. The standard design of the APR1000 has considered the types and magnitudes of the DBEH and RSEH that may occur in potential sites.

Design objectives against the external hazards are established as WENRA [IV-2] objective 2 (accidents without core melt) for DBEHs, which is 'no off-site radiological impact or only minor radiological impact', and objective 3 (accidents with core melt) for RSEHs, which is 'only limited protective measures in area and time needed for the public'. The objective 3 target is met if the design ensures the safety functions of items that are ultimately necessary for preventing early or large radioactive releases.

The conservative engineering design approaches applied in previous NPP designs are still effective for the DBEHs, while it needs to be demonstrated that the cliff edge effects are prevented using best estimate approach plus safety margins for the RSEH.

In the APR1000 standard design, in case the value of RSEH parameters is not significantly more severe than those of DBEHs (except for the earthquake, described below), a conservative design approach is taken, i.e. the SSCs required to be used for the RSEH are deterministically designed based on the RSEH parameter values. However, where the RSEH parameter values significantly exceed those of DBEHs, and thus the design based on those values is not viable, a probabilistic approach can be applied. For example, the temperature margin assessment approach evaluates the performance of the SSCs used to meet the objective 3 using realistic plant parameters under the RSEH air temperatures. Once the objective 3 target is achieved with a high level of confidence and with sufficient margin, then this process demonstrates that the plant is verified against RSEH conditions, and that cliff edge effects are avoided.

For the earthquake hazard, seismic margin assessment is used to probabilistically demonstrate that the high confidence level probability of failure for the containment is less than 1% to ensure that sufficient margin is provided for prevention of early or large releases due to RSEH earthquake.

The SSCs 'ultimately necessary for prevention of early or large releases' are identified based on the accident management strategies for DEC-A events initiated by the RSEH or DEC-B. The severe accident mitigation safety features, such as PECS, HMS, ERDS, diverse CSS, diverse SFP cooling system and the supporting HVAC, electrical and I&C systems and components together with the buildings that house those SSCs, were identified. These SSCs are classified as environmental conditions resistance level 2 and, thus, the corresponding environmental qualification and seismic qualification programmes are applied against the DBEH parameters followed by additional verification using seismic margin assessment and/or equipment survivability.

IV-5. NON-PERMANENT EQUIPMENT AND EXTENDED LOSS OF AC POWER

For the first 7 days after an AOO, DBA or DEC event, the APR1000 design does not require on-site or off-site NPE other than the off-site diesel fuel supply.

Safety systems and/or safety features for an AOO, DBA and DEC-A are designed to bring the plant to the safe state within 24 hours. After this time, the only concern is to maintain the long term safe state (or the containment heat removal) using the shutdown cooling system (or CSS). Since the diesel fuel oil storage for the EDGs and the alternate AC DG and the water inventory for the safety classified cooling tower basins are designed to last for 7 days, this safe state is maintained for (at least) 7 days without any NPE. However, additional off-site fuel supply for emergency power generation may be required after 7 days to keep the safe state. The raw water reservoir, which is designed to ensure for 30 days, enables use of the cooling towers from 7 to 30 days.

Similarly, the safety features used for DEC-B bring the plant to a SASS within 7 days after a severe accident. The only system required to maintain the long term SASS after 7 days is the diverse CSS and its supporting systems. No NPE is required for this operation, provided that additional diesel fuel supply is ensured from off-site.

Although the ELAP is considered as a beyond DEC-A event, it is considered in the design basis of the APR1000 for the robustness purpose and its safety objective is WENRA's objective 3.

Therefore, the ELAP management strategy of the APR1000 is established to maintain the core cooling up to 24 hours using the passive safety system of passive AFS, and it is assumed that additional long term cooling fails such that the core is melt down. After 24 hours, the severe accident mitigation safety features such as PECS, HMS, ERDS, diverse CSS and their supporting system safety features are activated using the power supply from the on-site mobile DG. Integrity of the containment is maintained such that early or large radioactive releases are prevented. The SFP water level can be maintained for the first 24 hours after ELAP without power, and then the diverse SFP cooling system powered by the mobile DG is used to maintain the coolability of the SFP so that the melting of the spent fuel assemblies is avoided.

The management of DEC-A events initiated by RSEHs is similar to that of ELAP, except that the use of permanent equipment of the alternate AC DG (instead of on-site mobile DG) is possible. The same safety features for the mitigation of severe accidents are used to maintain containment integrity. The cooling of spent fuel in the SFP is ensured by the diverse SFP cooling system, using power supply from the alternate AC DG.

IV-6. CONCLUSION

The APR1000 is designed to provide diverse safety features for DiD level 3b such as EBS, diverse CSS, diverse SFP cooling system and the associated cooling water systems, I&C systems, power supply and HVAC systems. In addition, it also provides safety features for the mitigation of severe accidents at DiD level 4 such as ERDS, PECS, PAR and diverse CSS. This design approach ensures a strong independence among different DiD levels.

The concept of 'practical elimination' of early or large radioactive releases is implemented by providing dedicated and/or robust design provisions, followed by demonstration using deterministic and/or probabilistic assessment. A quantitative probabilistic target is also determined for the demonstration of extremely unlikeliness of the sequences to be 'practically eliminated'.

The RSEH that exceeds the DBEH is considered in the design to ensure that the plant has sufficient safety margin to avoid early or large radioactive releases against these extreme natural hazards. The items that are ultimately necessary to achieve the WENRA objective 3 target against the RSEH are identified and designed to verify their functions against those RSEH parameter values.

NPE needed for all plant states, including the ELAP conditions, is identified on the basis of the corresponding accident management strategy and assessment. With the installation of robust diverse safety features designed to withstand the RSEH conditions, only a limited number of NPE is required for the APR1000.

REFERENCES TO ANNEX IV

[IV-1]   INTERNATIONAL ATOMIC ENERGY AGENCY, Safety of Nuclear Power Plants: Design, IAEA Safety Standards Series No. SSR-2/1 (Rev. 1), IAEA, Vienna (2016).

[IV-2]   WESTERN EUROPEAN NUCLEAR REGULATORS' ASSOCIATION, Safety of New NPP Designs, WENRA Reactor Harmonisation Working Group (2013).

[IV-3]   EUROPEAN UTILITY REQUIREMENTS ORGANISATION, European Utility Requirements for LWR Nuclear Power Plants, Revision E (2017).

[IV-4]   EUROPEAN ATOMIC ENERGY COMMUNITY, FOOD AND AGRICULTURE ORGANIZATION OF THE UNITED NATIONS, INTERNATIONAL ATOMIC ENERGY AGENCY, INTERNATIONAL LABOUR ORGANIZATION, INTERNATIONAL MARITIME ORGANIZATION, OECD NUCLEAR ENERGY AGENCY, PAN AMERICAN HEALTH ORGANIZATION, UNITED NATIONS ENVIRONMENT PROGRAMME, WORLD HEALTH ORGANIZATION, Fundamental Safety Principles, IAEA Safety Standards Series No. SF-1, IAEA, Vienna (2006).

LIST OF ABBREVIATIONS

ABWR	Hitachi-GE Advanced Boiling Water Reactor
AC	Alternating Current
AFCEN	French Association for Nuclear Steam Supply Systems Equipment Construction Rules
AFS	Auxiliary Feedwater System
AOO	Anticipated Operational Occurrence
AP1000	Advanced Passive 1000 MW
APR1000	Advanced Power Reactor 1000 MW
APR1400	Advanced Power Reactor 1400 MW
APR+	Advanced Power Reactor Plus
ASME B&PV	American Society of Mechanical Engineers Boiler & Pressure Vessel
ASN	Autorité de Sûreté Nucléaire, France
ATWS	Anticipated Transient Without Scram
CCF	Common Cause Failure
CCWS	Component Cooling Water System
CDF	Core Damage Frequency
CIS	Containment Isolation System
CNNC	China National Nuclear Corporation
CNNC'S HPR1000	Advanced Pressurized water Reactor 1000 (developed by the China National Nuclear Corporation)
CSS	Containment Spray System
DAS	Diverse Actuation System
DBA	Design Basis Accident
DBA-1	Design Basis Accident with frequency comprised between 10^{-4} and 10^{-2}/r.y
DBA-2	Design Basis Accident with frequency comprised between 10^{-6} and 10^{-4}/r.y
DBC	Design Basis Condition
DBEH	Design Basis External Hazard

DC	Direct Current
DCH	Direct Containment Heating
DEC	Design Extension Condition
DEC-A	Design Extension Conditions without significant fuel degradation
DEC-B	Design Extension Conditions with core melting
DG	Diesel Generator
DiD	Defence in Depth
EBS	Emergency Boration System (for APR1000); Extra Boration System (for EPR)
ECCS	Emergency Core Cooling System
EDG	Emergency Diesel Generator
ELAP	Extended Loss of Alternating Current Power
EPR	European Pressurized Water Reactor
EPRI	Electric Power Research Institute
ERDS	Emergency Reactor Depressurization System
ESF	Engineered Safety Feature
EU-APR	European-Advanced Power Reactor
FA3 EPR	EPR Flamanville Unit 3
FCI	Fuel-Coolant Interaction
FLSR	Flooder System of Reactor Building
FLSS	Flooder System of Specific Safety Facility
GPESPN	French advisory committee of experts for nuclear pressure equipment
GPR	French advisory committee of experts for nuclear reactors
HMS	Hydrogen Mitigation System
HPAC	High-Pressure Alternate Cooling System
HPME	High-Pressure Melt Ejection
HPR1000	Advanced Pressurized water Reactor 1000
HVAC	Heating, Ventilation and Air Conditioning

IAEA	International Atomic Energy Agency
I&C	Instrumentation and Control
IRSN	Institut de Radioprotection et de Sûreté Nucléaire, France
ISLOCA	Interfacing System Loss of Coolant Accident
IVMR	In-Vessel Melt Retention
LB-LOCA	Large Break Loss Of Coolant Accident
LCF	Late Containment Failure
LOCA	Loss Of Coolant Accident
LOOP	Loss Of Off-site Power
LRF	Large Release Frequency
MCCI	Molten Core Concrete Interaction
MSLB	Main Steam Line Break
NPE	Non-permanent Equipment
NPP	Nuclear Power Plant
NRA	Nuclear Regulation Authority, Japan
OPR1000	Optimized Power Reactor 1000 MW
PAR	Passive Autocatalytic Recombiner
PCS	Passive Containment heat removal System
PECS	Passive Ex-vessel corium retaining and Cooling System
PHRS	Passive Heat Removal System
PIE	Postulated Initiating Event
PRA	Probabilistic Risk Assessment
PRS	Passive Residual heat removal system of Secondary side
PSA	Probabilistic Safety Assessment
PWR	Pressurized Water Reactor
RCC-E	Design and Construction Rules for Electrical and I&C Systems and Equipment
RCC-M	Design and Construction Rules for the Mechanical Components of PWR Nuclear Islands

RCIC	Reactor Core Isolation Cooling System
RCS	Reactor Coolant System
RCW	Reactor Building Cooling Water System
R&D	Research and Development
RHR	Residual Heat Removal
RPV	Reactor Pressure Vessel
RSEH	Rare and Severe External Hazard
RSW	Reactor Building Service Water System
r.y	reactor year
SASS	Severe Accident Safe State
SB-LOCA	Small Break Loss Of Coolant Accident
SBO	Station Blackout
SFC	Single Failure Criterion
SFP	Spent Fuel Pool
SG	Steam Generator
SGTR	Steam Generator Tube Rupture
SIS	Safety Injection System
SRV	Safety Relief Valve
SSCs	Structures, Systems and Components
SSF	Specialized Safety Facility
STUK	Radiation and Nuclear Safety Authority, Finland
TEPCO	Tokyo Electric Power Company
UHS	Ultimate Heat Sink
U.S. NRC	United States Nuclear Regulatory Commission
WENRA	Western European Nuclear Regulators' Association
WWER	Water-Water Energetic Reactor

CONTRIBUTORS TO DRAFTING AND REVIEW

An, C.H.	KEPCO Engineering and Construction, Republic of Korea
Barbaud, J.Y.	Consultant, France
Bernard, M.	Electricité de France, France
Chitose, H.	Hitachi GE Nuclear Energy, Japan
Contri, P.	International Atomic Energy Agency
Hibino, K.	Nuclear Regulation Authority, Japan
Kang, S.H.	KEPCO Engineering and Construction, Republic of Korea
Kim, K.T.	Korea Institute of Nuclear Safety, Republic of Korea
Massara, S.	International Atomic Energy Agency
Morozov, V.	JCS Atomenergoproekt, Russian Federation
Nakajima, T.	Nuclear Regulation Authority, Japan
Nitheanandan, T.	Canadian Nuclear Safety Commission, Canada
Nolan, R.	Nuclear Regulatory Commission, United States of America
Suikkanen, P.	Radiation and Nuclear Safety Authority – STUK, Finland
Tiberi, V.	Institut de Radioprotection et de Sûreté Nucléaire, France
Yang, S.T.	Korea Hydro and Nuclear Power, Republic of Korea
Wang, C.C.	China Nuclear Power Engineering Co., Ltd., China

Technical Meeting

Vienna, 7–11 September 2020 (virtual)

Consultancy Meetings

Vienna, Austria, 12–15 March 2019

Vienna, Austria, 19–22 November 2019

ORDERING LOCALLY

IAEA priced publications may be purchased from the sources listed below or from major local booksellers.

Orders for unpriced publications should be made directly to the IAEA. The contact details are given at the end of this list.

NORTH AMERICA

Bernan / Rowman & Littlefield

15250 NBN Way, Blue Ridge Summit, PA 17214, USA
Telephone: +1 800 462 6420 • Fax: +1 800 338 4550

Email: orders@rowman.com • Web site: www.rowman.com/bernan

REST OF WORLD

Please contact your preferred local supplier, or our lead distributor:

Eurospan Group
Gray's Inn House
127 Clerkenwell Road
London EC1R 5DB
United Kingdom

Trade orders and enquiries:

Telephone: +44 (0)176 760 4972 • Fax: +44 (0)176 760 1640
Email: eurospan@turpin-distribution.com

Individual orders:

www.eurospanbookstore.com/iaea

For further information:

Telephone: +44 (0)207 240 0856 • Fax: +44 (0)207 379 0609
Email: info@eurospangroup.com • Web site: www.eurospangroup.com

Orders for both priced and unpriced publications may be addressed directly to:

Marketing and Sales Unit
International Atomic Energy Agency
Vienna International Centre, PO Box 100, 1400 Vienna, Austria
Telephone: +43 1 2600 22529 or 22530 • Fax: +43 1 26007 22529
Email: sales.publications@iaea.org • Web site: www.iaea.org/publications

23-01353E